PIT LANE
02
PITLANE
FIA FORMULA ONE WORLD CHAMPIONSHIP
GRAND PRIX
2001 FUJI
GRID ACCESS
SUZUKA
Round 17
TV / RADIO

F1 INSIDER

NOTES FROM THE PIT LANE

TED KRAVITZ

C CASSELL

For Jill and Robert,
with love and thanks for buying the papers.

First published in Great Britain in 2025 by Cassell, an imprint of
Octopus Publishing Group Ltd
Carmelite House
50 Victoria Embankment
London EC4Y 0DZ
www.octopusbooks.co.uk
www.octopusbooksusa.com

An Hachette UK Company
www.hachette.co.uk

The authorized representative in the EEA is Hachette Ireland,
8 Castlecourt Centre, Dublin 15, D15 XTP3, Ireland (email: info@hbgi.ie)

F1, FORMULA 1, GRAND PRIX and related marks are trade marks
of Formula One Licensing BV, a Formula 1 company.

Distributed in the US by Hachette Book Group
1290 Avenue of the Americas, 4th and 5th Floors, New York, NY 10104

Distributed in Canada by Canadian Manda Group
664 Annette St., Toronto, Ontario, Canada M6S 2C8

ISBN (Hardback): 978-1-78840-570-6
ISBN (Trade Paperback): 978-1-78840-571-3
eISBN (eBook): 978-1-78840-572-0

A CIP catalogue record for this book is available from the British Library.

Typeset in 11.25/16.5 pt Miller Text by Six Red Marbles UK, Thetford, Norfolk.

Printed and bound in Great Britain.

13 5 7 9 10 8 6 4 2

Publisher: Trevor Davies
Editor: Scarlet Furness
Copy Editor: Chris Stone
Creative Director: Mel Four
Picture Research Manager: Jennifer Veall
Senior Production Manager: Peter Hunt
Additional Text By: Adam Cooper

This FSC® label means that materials used for the product have been responsibly sourced.

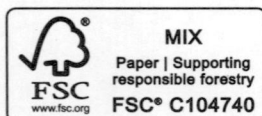

MIX
Paper | Supporting
responsible forestry
FSC® C104740
FSC
www.fsc.org

Contents

———

Introduction   1

Chapter 1: A Made-Up Job   5
Chapter 2: The Start Line   19
Chapter 3: The Murray and Martin Show   39
Chapter 4: Jacques, Gilles and Jerez   55
Chapter 5: Murray's Last Season   65
Chapter 6: Learning the Ropes   79
Chapter 7: Dealing with Disaster: The Pitfalls of
          Live Broadcasting   93
Chapter 8: Michael   103
Chapter 9: Nuts, Bolts, Bags and Flights   117
Chapter 10: The Six-Car Race   129
Chapter 11: From Ron to Ruin   143
Chapter 12: A Fifth and a Fix   163
Chapter 13: Brains and Brawn   183
Chapter 14: Schumacher's Second Coming   201
Chapter 15: The Piranha Club   211
Chapter 16: Sebastian   225

CONTENTS

Chapter 17: Nearly Men   239
Chapter 18: Ted's Notebook   253
Chapter 19: Surviving, Driving   269
Chapter 20: The 2021 Abu Dhabi GP   287
Chapter 21: The Eight-Pointed Stars   309
Chapter 22: The Finish Line   323

Index   333
Acknowledgements   341
About the Author and Picture Credits   345

The travelling circus

Introduction

'Hi, Ed.'

'Hi, Ted.'

I don't quite know how this has happened, but I'm walking down the pit lane of the Miami International Autodrome alongside singer-songwriting superstar Ed Sheeran.

'What do you make of Formula 1?' I ask him.

'It's nice, man. It's a circus, though.'

He's come a long way from South Suffolk to South Miami with many a packed stadium tour on the way, but even Ed Sheeran can see that F1 is a circus and in that, he's exactly right.

The Formula 1 world championship is essentially its own circus. It transports a show from town to town, performs to an excited audience, there's interest for the locals as the circus sprinkles its magic dust for a weekend, but come Sunday evening, it packs up and moves on.

The drivers and teams are the performers. Back in the day, Bernie Ecclestone was the ringmaster. Now it is the F1 organization, making sure the show starts and finishes on time. Media companies from around the world film and broadcast what happens in the ring. What actually happens in the performance – the result – is newsworthy, so journalists inform those who weren't fortunate enough to get a ticket for the big top as to what happened and why. There's always something going on, so for those there to report, observation, curiosity and an eye for detail are key attributes in order to fully understand the action and the movements behind the scenes.

It's been this way for 75 years in the form of the official Formula 1 world championship and for nearly 50 years of Grand Prix racing before that. And in each of those seasons, remarkable events have always happened. Stories that belong more in the theatre than a circus, often defying explanation as to how they came about. In searching for a reason as to how these stories – these legends – are made, I'm often reminded of a scene from the 1998 film *Shakespeare in Love*, featuring Geoffrey Rush as Philip Henslowe, owner of the Rose Theatre in London. Pursued by his patron Hugh Fennyman, played by Tom Wilkinson, Henslowe attempts to explain the conundrum of how the theatre business seeks to overcome its inherent 'insurmountable obstacles on the road to imminent disaster':

> Fennyman: 'So what do we do?'
> Henslowe: 'Nothing. Strangely enough it all turns out well.'
> Fennyman: 'How?'
> Henslowe: 'I don't know, it's a mystery.'

It's the same in Formula 1. Just as it looks like the championship is uncompetitive or the same team or driver is winning everything or the racing isn't as exciting as it used to be, along comes a Grand

Prix that has viewers jumping up and down around their living rooms, that has spectators trackside screaming encouragement at their heroes and even applause ringing out in the media centre. Sometimes it's caused by a brief rain shower, sometimes by a genius strategy, but these racing performances always rely on the skill and courage of the drivers. What particular combination of events got them to that point, nobody really knows – it's a mystery.

I've spent my professional life looking for the why's, the where's and the how's, piecing together little details that might be missed. Add them up, though, and you'll start to get a picture of what's *really* going on, and in this book I'm going to be sharing them with you. In these pages you'll be joining me behind the scenes in the pit lane and if you're curious to know how I do my job, consider this your handbook. But first, how did I get there?

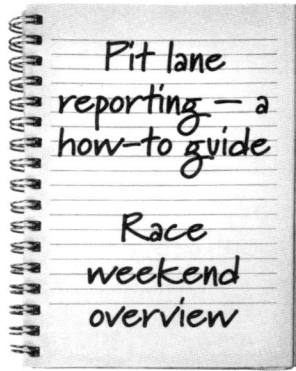

Pit lane
reporting — a
how-to guide

Race
weekend
overview

Chapter 1

A Made-Up Job

I'm standing in a gap between garages in the pit lane of the Suzuka circuit in Japan. It's raining, as usual. Mechanics from the McLaren Formula 1 team rush off the grid, pushing stacks of tyres into positions ready for the first pit stop. As I watch, there's a chug-chug from over my shoulder as members of the Sauber team force generator trolleys over a drain cover through to the paddock behind us. The trolleys are stuffed with bottles of compressed air, engine oil and tins of hydraulic fluid – crucial equipment on the grid, but useless to their cars once the race starts.

'Ah, here you are, ready to do your ridiculous, made-up job,' teases a passing Anthony Davidson, former World Endurance Car champion, Toyota Le Mans driver, and as an ex-Honda F1 driver, somewhat of a celebrity in Japan. This weekend, he's my Sky Sports F1 colleague.

'What do you mean?' I reply. 'I tried to do your job in Zandvoort,' he explains, 'and it was impossible. Nowhere to stand in the pit lane. Teams utterly unhelpful in giving out any information during the race. No one let me stand in their garage. Crap phone signal, way too much hiss through the headphones and I had to lug around that high-powered radio mic pack and battery you use. Unbelievable! I couldn't see a thing and didn't contribute to the commentary once. But here you are, and you've somehow managed to turn all that into a job. A job that you've basically made up! I don't know how you do it!'

Ah, Anthony, the job does exist, but what is it exactly? A pit lane reporter contributes information, explanation and observation into a race commentary from a position other than the commentary boxes. Usually this is around the garages in the pit lane, where race-critical pit stops take place and strategic decisions are made by the teams or the race officials. Having a roaming reporter able to go to where the stories are rather than being tied to a commentary box offers on-the-ground insight that, it is hoped, enhances the viewer's understanding and enjoyment of the race. It happens to be a pit lane in motorsport, but the job is similar to that of a reporter from a touchline, boundary, technical area or even from the 18th hole.

Building up to a golf, football or cricket event, reporters will often do 'stand ups' by the side of the playing area, ideally to inform viewers about an aspect of the field of play that will affect one side or the other. You can point something out on camera, in vision, but once the match, game or race gets going, a reporter will be a voice you hear over the main feed. In this respect, it's not dissimilar to a radio commentary, which is useful, as radio was my first professional love and the medium in which I'd planned a career before the lure of the F1 circus drew me in. As in any area of journalism, having a curious mind and the ability to understand and explain things are prerequisites for the job, but unusually in sport, the instrument with which the athlete performs in F1 is not as simple as a bat or racquet. It is a car made up of nearly 15,000 individual components, each playing a specific and vital role in keeping the thing going. Since Formula 1 cars cannot and do not carry any redundant parts – as the weight of these would cost performance – every bit has to fulfil its function, so it also helps to have an understanding of what those components do. In addition, an understanding of the physics of aerodynamics and the laws of mechanical engineering is desirable, although there will be ample on-the-job training opportunities to learn about aspects of F1 you had no clue even existed, usually five minutes before you need to sound like an expert on the subject in front of millions of people.

So, if that's what the pit reporter actually does, where do they do it? It starts with the first, second and third free practice sessions of an ordinary Grand Prix weekend, which you'll be spending walking up and down the pit lane. It is, after all, where all the garages are, where the engineers work, where the cars are built and repaired after the drivers have broken them, polished with the most expensive microfibre rags money can buy and then meticulously disassembled at the end of the weekend. It's also where most of the drama and intrigue happens, where plans are hatched, where snooping on rivals takes place and where drivers often lose their metaphorical rags with their competitors.

Additionally, the practice sessions have become part of the art of pit reporting, in that they are very useful for spotting new parts on cars. Upgrades, as we call them, have become a crucial part of telling the story of how a Formula 1 season evolves, explaining why the car that was once slow is now fast and why the car that was winning everything is now fighting for the minor placings. So much is learned and achieved by research and development over a season that an F1 team can often find significant seconds worth of lap time from March to December. Your best opportunity for identifying the bits that have achieved those performance gains is during the free practice sessions, which has made watching what is essentially 'training' a very popular part of F1 fans' weekend viewing. They also provide the commentators with an opportunity to familiarize themselves with each corner, the angles from which they are being filmed and what they look like on screen. These sessions are as crucial for the commentator and reporter as they are for the driver. This is your best opportunity to acquire essential knowledge and gather the thoughts that will help you read the story of the race to come.

The practice sessions also give us broadcasters the opportunity to get out more: to report from parts of the race track that will add insight into the prospective form of each car and to explain weird things that inevitably crop up, such as the trackside fires in the grass verges at the Japanese and Chinese Grands Prix, among others. The first time it happened in Shanghai we were surprised: there was no obvious reason for this seemingly spontaneous combustion. It was only when I walked down to the corner that I saw that there was a bump – invisible to the TV cameras – over which the cars were bottoming out, which was sending particles of burning titanium off the cars' skid blocks into the air, only for the wind to drift them on to the dry grass. In Japan, Sky Sports News's Craig Slater and I witnessed the marshals pre-emptively dousing the grass to make it less combustible, using anything that could hold water, from empty Bento boxes to upturned traffic cones.

While random trackside grass fires would be a good example of something that would call a halt to proceedings midway through, some sessions never get going in the first place, and the usual cause for this is the weather. F1 drivers are happy to practice, qualify and race in the rain, but there are complicated medical contingencies that need to be met if one of those cars were to crash and the driver to need transferring from the circuit medical centre to a more comprehensively equipped hospital. If the race track is in a city where the nearest hospital is only a mile or so away, such as Melbourne, Jeddah, Monaco, Monza or Singapore, then a regular road ambulance can be used. The problem is that most permanent race tracks are situated away from urban conurbations on account of the noise they tend to generate, so in the cases of Silverstone or Austria, Spa or the Nürburgring, a medical helicopter needs to be able to fly from the track to the nearest hospital and needs to be able to see where it is going. If there's low cloud, rain or fog, the helicopter cannot fly according to visual flight rules, and so to ensure nobody crashes their F1 car in the first place, a track session may be delayed indefinitely or even cancelled, leaving us on TV with some time to fill.

Time to fill is no cause for panic, as there are plenty of interesting things going on outside of the pit lane. You might head out to join the spectators in the grandstands, you might take the opportunity to talk to bored trackside marshals, or make the trek to one of my favourite spots to report from: the circuit helipad. It is almost always adjoining the medical centre, for obvious reasons. Why do I like it? Mostly because I like helicopters and talking to their pilots about the weather, and whether the practice session can start or not. Luckily the radio microphones and cameras we use are powerful enough to be able to broadcast live from these helipads, which I'm only too happy to do, more often than not brimming with the vital news that nothing continues to be happening.

With the practice sessions complete, or not, the next task for the pit reporter is qualifying. It's often the most exciting part of the weekend, where the cars are as light on fuel as possible in order to achieve the quickest lap time. The aim in qualifying is to deliver information into the commentary about any driver who has a car problem that is preventing them from getting out on track, keeping an eye on which compound of tyres everyone is using, and whether drivers are using a new set for best performance, or an already used set which means that their lap time won't be quite as good. Since there's more frantic activity in the pit lane than in the more relaxed practice sessions, reporters are required to stand in line with the front of the garages, on an actual, painted red line that has 'PIT LANE' helpfully added in white letters. You can move up and down the line, but only from the garage side – you're not allowed to step over it. From whichever part of the pit lane you choose, you can see what's coming out of three or four garages either side, but you can only see directly into the ones either side of you, unless you have a periscope on the end of a very long pole (not a bad idea, actually!).

Most drivers and engineers will tell you that qualifying is the most intensely pressurized part of the entire race weekend, and they're only too happy when it's all over, and they've not made some huge and costly mistake.

Once a driver has qualified on pole, the race is a doddle. OK, that's not quite true, as the current iteration of the rules demands that at least two different tyre compounds are used. Unless there is a race stoppage which allows everyone to change their tyres when waiting for the resumption (as happened in the 2024 Monaco GP), every car will have to make at least one in-race pit stop. Obviously, a load of time is lost having to slow down to come into the pit lane, have your tyres changed by 20 motivated mechanics in two seconds flat, and then get going again, and therefore drivers try to make as few pit stops as possible. But sometimes the tyre wear makes the car so slow that it's actually quicker to make another pit stop to take on fresh rubber for the race to the end. Who does what in these situations is called race strategy, which for obvious reasons the teams want to keep secret. A fundamental part of the pit lane reporter's job, therefore, is trying to figure out what they're up to and reporting back to explain it during the race.

It's also my job to tell the commentators if one of those pit stops has been particularly slow – or fast – and how that might affect the remainder of that driver's race. The key is to provide the viewers with as much relevant information as possible. If you approach the pit lane reporter's job that way, you won't go far wrong.

What complicates matters is that at most circuits I can't actually watch the race when I'm in the pits, and unless there's a big screen situated above the start/finish line grandstand, I have no way of seeing the race coverage. Now I realize this might seem odd for someone whose job is literally to tell people what happened in that day's Grand Prix, but it's true. At the majority of events the first time I watch qualifying or the race will be when I get home on Sunday night or Monday morning, or if I manage to catch any snippets on social media on the bus to the airport.

There's a very good reason why this is not problematic. Given that the pit reporter is there to tell the commentators what they *can't* see on their TV screens, there is no point in watching the same race coverage and looking at the same timing screens as the commentators – they've already got that covered. At the more modern circuits like Yas Marina in Abu Dhabi there's a big screen temptingly visible from the pits, but beware – if you end up just watching the race like everybody else, you'll be missing the stories the cameras don't see.

Even without a big screen to watch it's a mistake to stay put in the same location for the whole race. Going from garage to garage, talking to your contacts in the teams directly rather than by text message, will always result in quicker, better, fresher information that will be of value to the listener. Texts or WhatsApp groups can be an effective way for teams to keep every broadcaster and journalist informed at the same time, but where possible it's always better to get that information face to face. Plus, if you're the only reporter consistently going from garage to garage, as I seem to have been over the last 23 years, you'll find you tend to get the story before anyone else.

The first job is getting past the security guard at the back of the garage, and you're only going to do that if you have been granted special permission to watch from there. It might surprise you to learn that some of the most secretive and guarded teams are not the frontrunners – they're actually the teams whose cars are the slowest, which makes me smile, when I encounter their stonewalling, and wonder what it is they're so anxious to protect. Mercedes, McLaren and Ferrari, on the other hand, are comparatively open. The teams employ press officers or, as they're known more generally, communications managers – effectively media liaisons – to deal with enquiries from the written and broadcast media and it's worth remembering that their aim at all times is to present their teams in the best possible light. Protocol demands that rather than talking directly to a mechanic, engineer or technical director during the race, I consult the press officers first. I'm trying to keep across what's going on with 20 drivers and how they're all doing in the race – the fact that team press officers are only interested in their two drivers is helpful because they can fill me in on anything I might have missed while I've been covering other cars. They will always have a line about what the intention is for their driver in this stint, or explain whether they are saving tyres, engine or going flat-out.

It's pretty rare that you go into a garage and you can instantly see that something's wrong. So what are you looking for? The most obvious clue something isn't right is when you see a technical director leave the pit wall and cross the pit lane to the garage. If they go back and forth, you'll know that something's up with one of the cars.

Let's say you've done your work beforehand and convinced a team press officer to let you into their garage for the race. Well done, but you then have to be mindful about what you're reporting. For example, back in the days when we had refuelling I could sometimes see the figure revealing how many kilos of fuel they were about to put in the car, but obviously I couldn't broadcast that as it would be tactically useful for rival teams. So there's always an element of give and take.

If I can't see the pictures, how do I follow the race? By listening to it. I'm always listening to the commentary, effectively treating every word with the value a radio listener does, concentrating on the descriptions in the absence of visuals. And as a bit of an aid, I carry a fairly powerful monocular (basically a telescope with a wide field of vision) in my back pocket that is brilliant for seeing details far away down the pit lane. For example, if I'm watching from a midfield garage, and a Red Bull or McLaren comes in with a problem at the far end of the pit lane, with the trusty monocular I can see if there's a mechanic topping up the engine air system, or removing some debris from a radiator duct. Very useful, that little telescope. In addition, I have the live timing app running on my iPhone, which is really all you need to read a race without pictures. It gives race positions, lap times, gaps to the leader, intervals between cars, and sector times and speeds.

Once the race is over and the last car has returned to the area of the pits reserved for technical checks, reporters are allowed into the pit lane proper to grab some interviews with the team principals or top engineers as they come down from their pit wall gantries. This is always a highly charged and emotional time for the team personnel. You, meanwhile, need an interview, and so you need to pick your moment carefully. I've seen many race victories, championship wins, near misses, race losses and the aftermaths of accidents from here. Everyone deals with triumphs and tragedies differently, and you have to know how to treat each team boss.

McLaren's Ron Dennis was, for all his vaunted self-discipline and self-control, often surprisingly emotional when he came off the pit wall. I've experienced him both with eyes filled with tears, and lip curling with anger. By contrast, Ross Brawn (Ferrari, Brawn Grand Prix and Mercedes) kept his emotions firmly in check. He would give his colleagues a handshake, briefly look through some data on the pit wall, and then turn in my direction to grant an interview, usually, I suspect, because he wanted to get it out of the way in order to go and discuss things in further depth with his engineers. Whether his team had done something incredibly good, like win a championship, or something incredibly bad, like ordering Rubens Barrichello to give up the win of the 2002 Austrian GP, Ross was always as close to being an emotional flatline as I've seen. The exception was when, having saved the former Honda team from going out of business, he won both the 2009 drivers' and constructors' titles with a car bearing his name. On that memorable afternoon in Brazil Ross was in pieces.

To get the interviews you need, it's vital to learn everyone's preferences. Ferrari's Fred Vasseur leaves the pit wall extremely quickly, sometimes before the chequered flag has even fallen. By the time you get to the Ferrari pit, he's long gone, usually across to the garage, or back to his office, so you can only ever catch up with him later on in the paddock. One of his predecessors at Ferrari, Stefano Domenicali, now the boss of the F1 organization, had a set routine, of which his interview with me was a part. He would congratulate or have a chat with colleagues on the pit wall, then go into the garage and shake hands with his mechanics. He'd then come out front to grant an interview that just happened to have the Ferrari garage and its sponsors' logos visible in the background. Clever man!

Once you know your team bosses and they know you, they become the easiest interviews to get. It's almost like they're expecting you to be there, and they're disappointed if you're not. The hardest interviews tend to be when the team bosses have had a very bad day, when their drivers have collided with each other, or they've lost a race or a championship. Or even worse, when it's the team that is at fault, rather than the driver. In those instances it's very rare that a two foot forward 'sticking the boot in' question will yield the best results. I'm all for being direct if the situation demands it, but most of the time, you'll get a worse interview if you go in with a 'Why on earth did you do that?' approach. In general, a bit of empathy and a sympathetic request for an explanation will almost always yield a more informative answer for the viewer.

Sometimes bosses can go into defensive mode if they anticipate they're going to be under attack from the fans or the media, whereas others are less guarded. In 2016 Mercedes teammates Lewis Hamilton and Nico Rosberg collided on the first lap of the Spanish GP. In the immediate aftermath I found F1 legend Niki Lauda, who since 2012 had been attached to the Mercedes team in the chairman role, pacing around outside the back of the garage, clearly furious. He wasted no time in telling me how angry he was with the drivers. In his view it was completely unacceptable. One had forced the other on to the grass, only for the other one to turn in, and make sure the two crashed. By contrast, after the race, in damage-limitation mode, team principal Toto Wolff was more measured, arguing there was blame on both sides, that the drivers were generally free to race but that he would be having words with them.

These post-race interviews are key to understanding the outcome of the Grand Prix. They often reveal newsworthy stories that change the viewers' perception of the events they've just spent a couple of hours watching. It's my job to ask the questions that I believe our viewers at home would ask if they were in the pit lane at that moment.

I'm yet to mention two final aspects of the job, namely features, as we call them – three- or four-minute pre-made video stories that run pre- and post-race, and the notebook, my pit-lane roundup that goes out live after qualifying and the race. Features take up a disproportionate amount of our time over a weekend, because they're highly crafted, tightly edited pieces of work. They are devised mainly by our production team and myself, before being scripted, filmed and edited, just in time for transmission before the race.

And finally, the aspect of the pit reporter's job I did truly invent: *Ted's Notebook*. The essential ethos is simple. It's one shot. One take. Once we start, we're going out live, and we don't stop, so anything can happen and often does. It came about because I realized so many of the stories about the drivers who were not the frontrunners were getting lost. And so the purpose of the notebook is to tell the story of every driver's day.

So there you go, Anthony Davidson, that's my made-up job. But it's quite a fun one, and if we can make it feel like a little race club, with all of us rev-heads watching together based upon our shared interest in F1, a club that anyone can be a part of and feel included in, then it won't seem quite so ridiculous.

Chapter 2

The Start Line

———

Strangely for someone who is terrible at maths, I find comfort in numbers. I was one of those children who found that engaging with them helped me make sense of things I didn't fully understand. I found numbers pleasing to look at and to write, I knew where I was with them. If numbers represent order in the world, F1 is full of numbers and, growing up, I found that very reassuring. Even today, in an anxiety-inducing world, I find much comfort in a neat table of motor-racing finishing positions, lap times and points scored.

The rest of my family are literary and artistic types. My father ran an agency representing classical musicians, conductors and opera singers, while my mother worked for many years in book publishing, and – so that they could keep across the reviews – we always had stacks of newspapers around the house. After my mother had taken the books section, my father the music reviews and my eldest brother exercising his right as firstborn to the TV guide, I was left with the news, politics and sports pages. And because my middle brother had control of the TV remote and wasn't interested in my preferences, I spent my free time after school reading the papers on our kitchen table. Those hours must have left their mark, because I came to appreciate the way the news was reported, the prominence that certain stories were given, and the skill of the journalists in telling the reader why they should care, all of which gave me the first sense that I might enjoy a career in journalism.

Reading and enjoying the political coverage in newspapers like *The Times*, *The Daily Telegraph*, *The Independent* or *The Guardian* set me up for my university degree in politics. But the stories I looked forward to reading most were to be found in the sport section. I was interested enough in the three or four pages of football news and results, and as you now know, I was very happy reading the numbers in a league table – matches played, won, lost, drawn, goal difference and points – and if a big tournament or championship was on, I loved reading about the competitors' stories. But what I really wanted to read about was Formula 1. F1 coverage in the newspapers was pretty hit and miss when I was growing up: if you wanted to read those stories you had to hunt for them. And so that's what I did.

The race or qualifying reports were always welcome, but what I wanted most were the numbers. Growing up pre-internet and even pre-regular TV coverage, the only place you could access the full lap-time qualifying results or race classifications was the following day in the newspapers. Happily my parents would generally have a copy of every Sunday paper and thus Sunday mornings would find me forensically examining the lap times from the previous day's qualifying and building my own story of what had happened – who did well, who did badly, what the gaps between the drivers were, and what stories the lap times told.

It's not really any different from what I do today. Much as I love numbers, I'm not great at remembering them. *Ted's Notebook* was partly born from the notebook I always have in hand where I've recorded numbers to the thousandth of a second.

As a kid at that kitchen table I found that F1 had a curious pull on me – the more I read, the more I wanted to read. The more I learned about these teams and their drivers, the more I wanted to learn. It got to the point where I would scour every newspaper every day to see if they had an F1 story buried in the back pages, and I'd devour even the smallest article. Thursday morning's trip to the newsagent to pick up a fresh copy of *Autosport* magazine was a weekly highlight. My family weren't the least bit interested in F1, and were often heard to wonder where this fascination had come from.

Looking back, it was probably Nigel Mansell's exploits in the mid-eighties that hooked me in. I particularly remember watching the 1987 British GP on BBC television, and Nigel's famous charge to overtake and beat his nemesis and Williams teammate Nelson Piquet. Drawn to Nigel as I was (at the time a British household name), I found myself more intrigued by drivers from faraway lands with interesting names and flamboyant characters. As I grew older the Brazilian driver Ayrton Senna started winning races regularly, and I realized I'd found my racing idol. His skill, his flair, his handsome features, the fact he was so good, the best of his era, all that pulled me in. I watched all the races. Whether they were broadcast live into *Sunday Grandstand* or were just a 45-minute highlights show, I would be glued to the screen. Murray Walker became the soundtrack of my weekends, with James Hunt and then Jonathan Palmer alongside the great man.

When I turned 17 and learned to drive, I became even more interested in the physics of forward motion, of acceleration, of braking, the difference between understeer and oversteer, or, in the case of my Peugeot 106, chronic understeer. Knowing what 50 or 70mph actually felt like from a driver's perspective only deepened my motor-racing obsession.

I could never persuade any of my family or friends to go with me to a Grand Prix, and even back then it wasn't affordable. The first time that I saw F1 cars on track was in 1992 at an *Evening Standard*-sponsored racing day at Brands Hatch, free to attend with the voucher in the newspaper. Brands Hatch was the closest circuit to my home in London and I was old enough to get there by myself on the train, voucher in hand. The star attraction among the Minis and Formula Fords was a special appearance by Team Lotus. Johnny Herbert was the home hero, and his enthusiastic fans followed him around the paddock chasing for an autograph. His teammate Mika Häkkinen was quietly keeping to the sidelines. When I spotted him, I was quite excited. I was probably one of the few people there who knew who Mika was, as I had been keeping an eye on his early career. I went to meet him. He was sitting on a tool chest handing out autograph picture cards to those who'd been disappointed not to meet Johnny, although he hadn't actually signed any. He was as taciturn then as he would go on to be in his future racing career, although as we would later find out, his emotions ran deep.

The highlight of the day was a set of laps from the Lotus-Ford F1 car, and when Johnny booted it out of the last corner, Clark Curve, the acceleration and the sound of that engine brought a huge smile to my face. To experience the drama of an F1 car on track with the roar of the V8 engine was like nothing I'd ever encountered. By now, F1 was firmly my life's passion. Not for one second, though, did I ever think that I would work in it. I just didn't see a way in. And at this point, I was focused on a more realistic career prospect, radio journalism.

My fondness for radio developed in my teens. I was a bit of a night owl, and I loved the immediacy and intimacy of a radio programme – the thought that there was someone else on the other end of an airwave, awake at the same time, making us listeners feel we were not alone. In the late eighties and early nineties nobody did that better than LBC's *Through the Night* show.

The programme, on-air from 1am to 4am, was presented most often by Clive Bull, who created a society of listeners. His regular listeners might not have had the same interests or opinions, but what we did have in common was that we were all awake at that time of night, some working, some not, but all listening out for each other.

As much counsellor, lawyer and therapist as he was the presenter, Bull usually devoted the first hour to 'calls on the topic of your choice', which allowed regulars like Babs from Bermondsey or Charles from Camden to phone in with updates on their lives. As members of the club, fellow listeners were always interested to hear their stories. Clive would introduce segments like 'guess the mystery noise', which was usually made by some office accessory that he had picked up in the deserted LBC newsroom, or 'the rolling quiz', where a caller would set a question and the first person to answer it correctly would then set the next question, and so on, carrying on throughout the night. These kept me listening into the early hours, the other hook being that you never knew who was going to call up next, and neither did Clive. Comedian, writer, actor and author Peter Cook was a regular caller under the pseudonym Sven from Swiss Cottage, purporting to be a visiting Norwegian whose wife Jutta had walked out on him and was at large somewhere in Continental Europe. Sadly, Sven never found her, instead distracting himself with late-night complaints to Clive about the national obsession with fish in his native Norway. I recommend a documentary made for the UK's Channel 4, easily found online, called *Night Caller*, all about Clive Bull's *Through The Night*. It's well worth your time.

Radio would eventually lead me to Formula 1, but it was Exeter University that led me to working in radio. I had chosen to study politics, a subject I found pretty interesting and, crucially, pretty easy to pass exams in. Thus, A-Levels followed by a degree in political science made sense. Exeter was a fine place to do a politics degree, but my real interest in applying there was the student radio station, University Radio Exeter, or URE for short – at that time one of the best in the country. My course only required four hours of lectures and four tutorials per week, so I spent the rest of my three years there learning how to present and produce radio, while spending my Sunday afternoons in the company of Steve Rider and Murray Walker (and thanks should be offered at this point to the patience of my fellow Mardon Hall residents who were kind enough not to mind too much that I regularly commandeered the shared TV room on Sunday afternoons).

While at university I also learned the hard way how difficult it was to make any money in motorsport when I set up the extremely unsuccessful and ultimately loss-making Exeter University Motorsport Club. The plan was that like-minded petrolheads and

I would watch the races together in the student union bar and then have a go ourselves at a karting track that had been set up inside Exeter's Westpoint Arena. Unfortunately, I got my finances wrong, and having paid to hire the karting venue at a fee of around £500, discovered that only six people had showed up for the kart race, and that they couldn't afford to pay more than a tenner each. The sponsorship income for untalented university students being non-existent, I ended up funding the rest of the track-hire fee from my savings.

University Radio Exeter was much better run and maintained high standards, so full training in microphone technique, editing, reporting, writing and general presenting was supplied by the older students and by John Whitworth, a radio professional known affectionately as 'Frog'. Once you had learned the ropes there was an informal talent-scouting relationship with the professional local radio stations, who were open to finding new voices for freelance reporting or presenting on those shifts at times when nobody else wanted to work. I never really felt like a BBC kind of person (although much later I did end up working for the corporation for three happy years), so I focused on trying to get in the door at the local commercial station, Gemini FM. A meeting with news editor Marcus White led to a couple of try-outs and then a shift reporting on *The Devon County Show*. But before my voice could appear on-air Marcus wanted a word about my name.

'Right, Ted, what's your surname again?'

'Slotover,' I replied.

'Well, we can't call you Ted Slotover, it's too awkward for radio. Have you got any other names?'

I should mention at this point that Marcus White had a lisp, and he may have had more interest than most in avoiding names beginning with S.

I thought for a second and said, 'Well, my mother's maiden name is Kravitz.'

He sounded it out a few times. 'Kravitz, Kravitz. Ted Kravitz. That'll do!'

My true passport name is no secret, but professionally I've taken my mother's maiden name ever since that day. There wasn't any great discussion about it. I wasn't going to invent a cool name, this happened to be the only other name I had. My maternal grandfather, Richard Kravitz, was delighted. He was in the media world, publishing *Esquire* in the UK and *Boxing News*. In fact, I used to go with my uncle Pete to sell copies of *Boxing News* outside the York Hall in Bethnal Green when there were fights on.

With some on-air reporting for Gemini FM under my belt, I now had some demo tapes with which to try and secure more work in what was increasingly looking like my chosen career. The Radio Academy ran a training weekend with an opportunity to submit a half-hour radio documentary that would be judged by a panel. I'd pulled an all-nighter to get mine edited. The next day's talk was being given by one of the heaviest hitters in radio broadcasting, Richard Park. At the time Richard was programme director at London's Capital Radio, then the most popular and successful commercial radio station in Europe. Ordinarily I would have been too terrified to approach him, but with the usual alerts in my sleep-deprived state not quite working perfectly, I found myself walking up to him. I introduced myself and told him that I'd been working in Devon but lived in London, and that I really wanted to work at Capital Radio. Was there anybody I could talk to at their Euston Tower base about some shifts?

Sizing me up (my decent voice, the fact that I'd gone up to him to ask for a job), Park gave me the phone numbers of the news editors at Capital, David Hedges and Patrick Johnston. As a regular listener to 95.8 Capital FM (Capital's on-air name), I already knew of Patrick. He was a superb news presenter and, together with Hedges, delivered it in the distinctive and accessible style that was proving so popular as an alternative to the traditional BBC news offering.

Hedges had a few early breakfast news-reading shifts that, for obvious reasons, his usual pool of reporters weren't that keen on filling. He agreed to try me out on a few to see if I was up to it. Early breakfast sounds quite dynamic, but really, it was a night shift. I had to wake up at 3am to be in the newsroom on the ground floor of the Euston Tower by 4am. I would then read the news wires, look at the morning's papers (the first editions of which were already strewn around the newsroom) and from those, write a one minute to 90-seconds long bulletin, and then read it out on the air at 5am. There was no time for nerves or to think about how many people were listening. The 5.30am headlines came up quickly, and then it was on to the 6am and the 6.30am slots, before writing another updated bulletin for the main breakfast newsreader.

All breakfast shows are the most listened-to programmes on their radio stations. At a time when Capital Radio was the biggest commercial radio station in Europe, *Chris Tarrant in the Morning* was its headline show. We were all a bit in awe of Chris, as he really was a very famous household name and absolutely at the top of his game. More approachable was Tarrant's newsreader, a wonderful man with a booming voice called Howard Hughes, for whom I wrote those 90-second bulletins at 6.50am each day.

Howard was a radio nut. He loved all things radio and was blessed with one of the best voices of all time. He was also very, very funny and had a few highly individual quirks, one of which was reciting his own, unrepeatable, voiceover introduction to the ITV news. Another was his affection for an aftershave called Jazz. It came in a bottle with a black and white cap, and Howard would always spray about five squirts around his neck before he left the newsroom to go on-air. Was it that the alcohol in the aftershave helped his voice sound more booming and resonant? Or did he just really like the scent?

Every day Howard would burst into the office at 6.50am. 'Ted, good morning!' he would boom. 'What are you leading with today?' Most of the time I had what he considered the correct top story, or he would say, 'That's fine, I'll just change this a bit', before going into the studio and reading the 7am news with small alterations to the script that I'd prepared. One particular day there had been a storm overnight in London, and as I was driving in at 3am I had seen that some scaffolding had come down and fallen billboards and other debris were lying around the streets of the West End. This seemed to me important news for listeners on their way to work that morning, so I made that the top story. It was the one time Howard wasn't happy with my choice. He didn't exactly say, 'This is weather, not news,' he just said, 'Hmm, interesting choice. No, I don't think we'll lead with that, I think we'll lead with another story today.'

Gradually, through moments like that, I was learning about telling a story, painting a picture, the importance of headlines and the priority that stories should get – the nuts and bolts of editorial.

Something else Howard understood was about the times that people would be listening to the radio. Bear in mind this was the mid-nineties, when the first thing people did to find out what was happening in the world when they woke up in the morning was turn on the radio, rather than picking up their phones.

'What's the point in doing this early breakfast news at 5am?' I once asked. 'Why don't we just take the feed from Independent Radio News?' Howard replied, 'There are more people listening to Capital FM at 5am than there are at 5pm, and that's why it's important to read our own news.' Hard to believe, but the listening figures showed he was right.

I loved it at Capital. The Euston Tower and Leicester Square studios were always exciting places to be. I thought that I was going to carry on working in radio. I enjoyed doing it, and I'm blessed with a decent enough voice. I still loved F1, but there was no real opportunity to work in the sport as far as I could see. It was on the BBC, Murray Walker did the commentary and that was that. It wasn't even a realistic dream, or something I ever thought I could get into. That all changed on an otherwise ordinary day in September 1996.

I was coming to the end of my early breakfast shift, which included a handover meeting with the day newsreaders and a planning meeting for that evening's edition of Capital's news show *The Way It Is*. An alert popped up on the Press Association's planning schedule: 'Arrows F1 team press conference at Chelsea Harbour.' The rest of the newsroom had never heard of Arrows. The sports team had, but even they weren't particularly interested.

So with my metaphorical F1 anorak on, I explained that this was going to be an announcement that soon-to-be world champion and household name Damon Hill was going to be driving for these

Arrows chaps next season, team boss Tom Walkinshaw having convinced him to join what was then a back-of-the-grid team. My editor wasn't aware Damon had been dropped by his current team, Williams, despite the fact that he was about to win them the championship, and he cared even less about Arrows. However, I managed to convince him that even though my shift had finished, I should go and record the audio from the press conference, just in case they did want to run a clip that evening. With my Capital FM press card, and a tape recorder and microphone, I hopped on the tube down to Chelsea Harbour.

Damon looked understandably miffed at the whole situation. Most likely his mind was on the world championship-deciding race in Japan, which was still a fortnight away. His manager Michael Breen was there, looking rather happier than his client, probably because the multi-million-pound contract with Arrows would result in a nice commission for him. All the F1 correspondents from the newspapers I had scoured for F1 news for so long were present, including Derick Allsop (*The Independent*), Stan Piecha (*The Sun*), Ray Matts (*Daily Mail*) and Bob McKenzie (*Daily Express*). They asked Damon questions along the lines of 'Why Arrows?' and more forthrightly 'You know you're not going to be winning any races with Arrows, how much of a come down is this for a guy who looks like he's about to become the 1996 world champion?' Damon went from looking a bit miffed to increasingly pissed off as the press conference went on.

While Damon grappled with the implications of his team move, there was also a significant change going on in the world of F1 broadcasting in the UK. After 20-odd years of showing occasional qualifying sessions and most of the races to devoted viewers such as myself, the BBC had lost the rights to televise F1 to the ITV network, who were preparing to appoint a production company to make the programmes for them. I didn't know it at the time, but I was about to meet someone at that press conference who would change my life. I had been aware of James Allen through his news editorship of *Autosport* magazine and as the presenter/reporter on *Nigel Mansell's IndyCar 1993*, a programme I and many F1 fans had enjoyed, documenting Nigel's championship-winning season with Newman-Haas in the USA. That programme had been made by a production company called Chrysalis Sport, and I guessed that James might also be involved in their bid to produce ITV's F1 coverage for the 1997 season.

I spotted him watching the press conference scene from the back of the room, so afterwards I went up and introduced myself. We fell to chatting about the Chrysalis Sport bid and I asked if they might be looking for someone junior who could tell Jacques Villeneuve's helmet from Damon Hill's. James smiled. He couldn't speak for Chrysalis, but what he did do was tell me who to write to (and I do mean a written letter – email was in its infancy in 1996). This was the producers Rupert Bush and Dave Lewis. I wrote, and wrote again, then phoned to follow up, and then phoned again. If only to stop me bothering them, Dave and Rupert finally took my call and invited me into their office, a converted church on Camden Park Road, for a chat. In fact, were it not for the patience and kindness of Valerie Garford and Sarah Needham, who had to field my many calls to the Chrysalis Sport producers, I might never have got through the door.

As a trial task Dave and Rupert gave me a race to log. When you're editing a video feature you need to have footage – clips – to cut together. Before searchable digital video libraries, the only way to know what shots were where was for somebody to physically watch all the tapes and log it down on pieces of paper (which you then slotted into the tape box, or a central file, so that anyone could look through it and find the shots they needed). A tape-logger was a good place to start for a junior and was the way lots of people found their way into the television industry.

I turned out to be more than competent at watching back the races and writing a description of the shots and the audio or commentary with timecodes, and Dave and Rupert told me to stay in touch. At that point they were waiting to find out if they had won the production contract, or if it had gone, as most TV sport industry insiders predicted it would, to the in-house production arm of Carlton, ITV franchise licence holders for the London region. Carlton had already lined up BBC legend Murray Walker to continue to be lead commentator for ITV, and had chosen Jim Rosenthal to present. I never found out if Carlton was aware, but Chrysalis had also signed up Murray. Smart operator as he was, I think Murray had told every bidder that he was on their team! Chrysalis (whose bid partnered up with the Meridian and Anglia ITV franchises, creating the neat production name MACh 1) had Steve Rider – seasoned BBC sports presenter, who anchored prestige shows such as *Sportsnight* and *Grandstand* – optioned to present the ITV F1 show if they won the contract.

And much to everyone's surprise, they did. Suddenly MACh 1 had to organize one of the hardest jobs in TV sport – live, on-site presentation of F1 on a grander scale than had previously been seen on the BBC. Arguably, Chrysalis had won the bid because they convinced ITV they could do it bigger and better than anyone else.

Now they had to deliver. I got the call I had been hoping for, and was offered a job to join the production team as junior researcher. I met the top boss Neil Duncanson and the rest of the team: field producer Alan Hurndall, VT producer John 'Noz' Nolan and production managers Tim Breadin and Karen Raphael. A small group, now about to produce a big sport.

There was so much to do in just a few months that I didn't have time to think about the fact I was now working in F1. I spent the winter of 1996–97 in a darkened room with a TV and a tape machine logging all the tapes of the previous season's racing that Chrysalis had received from Formula One Management, the company headed at the time by Bernie Ecclestone, that controls the commercial and broadcast rights of Formula 1. I was in heaven – the opportunity to come into work, sit down in front of a TV, put on tapes of races and write down what I was seeing was bliss. I couldn't believe that I was being paid for it!

Chrysalis had some problems to solve. Earmarked for the presenter's job, Steve Rider decided that he wasn't quite ready to leave the BBC just yet. As he told me many years later, when we worked together, at the time he'd had the feeling that the senior executives at ITV weren't particularly keen on bringing in a major signing from a rival network, preferring to use their own people. Chrysalis boss Neil Duncanson pushed back against that, arguing that F1 needed a fresh start with a knowledgeable presenter, rather than an existing ITV name brought in from football. But for now, Rider had decided to stay at the BBC, and if he was out, who could do it?

So it was that, in November 1996, I found myself at Molinare Studios in London's Carnaby Street, helping out at a screen test to find the UK's next F1 presenter. The three candidates were all racing fans: Mike Smith, a well-known TV presenter and sometime racing driver; David Smith, who used to read the news on Channel 4's *Big Breakfast* ; and David 'Kid' Jensen, the Radio 1 DJ. Our setup was basic to say the least. I set up a table in a small studio with some tablecloths and a chequered flag, and the three candidates did their screen tests. Their task was to introduce the viewers to the programme, perform some links, read out the world championship standings and throw to James Allen, signed-on with Chrysalis as pit lane reporter, who would be on the other side of the studio with a camera and microphone, ready to do a pretend report.

As the day wore on James suggested we move out of the building, so I found myself in Foubert's Place, looking after the cameraman's back and the cable that trailed into the studio. While James was doing his mock pit report a dog came up and peed against a lamp post. 'This could mean trouble,' James said, without missing a beat, 'the car has sprung a leak,' to much hilarity in the gallery.

The three presenters were all very good. However, in the end ITV stepped in with their choice of Jim Rosenthal, who was their well-known main anchor for boxing and a presenter on some football programmes. As noted, Carlton had also seen Jim as the man for the job; he had been signed up to present for them had they won the bid. Being a football man at heart, Jim was always quite honest about the fact that he was not an F1 aficionado. He kept up with the main stories, but he wasn't an expert and didn't pretend to be. In fact, it's not essential that your main presenter has that expertise: that's what your guests and reporters are for. You need the presenter to be a rock-solid steady pair of hands to keep a live, three-hour show running – no easy task – and know to ask the pundits the right questions.

The year 1997 started with a pre-season press launch outside ITV's headquarters in Gray's Inn Road, unveiling the full presentation team: Murray Walker and ex-F1 driver Martin Brundle commentating, Jim Rosenthal anchoring, with Tony Jardine (who had previously worked in motorsport PR) and Simon Taylor (writer and columnist) as expert analysts, plus James Allen and Louise Goodman (formerly Eddie Jordan's press officer at Jordan Grand Prix) as reporters. I remember how delighted Murray was to be there, knowing he would still be commentating on the sport he adored. As I later learned Martin had more doubts, feeling uncomfortable moving into a TV role when he still wanted to be driving. Thinking he'd been pushed out of F1 too soon by Eddie Jordan, for whom he raced in 1996, Martin hadn't given up, and continued to hunt for an F1 drive while he was commentating throughout '97.

It was a pleasant enough press conference. The main points of interest seemed to be that Murray would be continuing as the voice of Formula 1, and Louise's groundbreaking role as the first full-time female F1 reporter. More unwelcome were the headlines the next day, which focused on the negative impact for F1 viewers of the races being shown on a commercial channel. It had been revealed that ITV would have to squeeze five two-minute advert breaks into each Grand Prix. For ITV, it wasn't negotiable. A commercial channel makes money from putting ad breaks in their programmes, that's how it works. But while a 45-minute half of a football match is just about long enough to air without an advert break, a 90-minute to two-hour Grand Prix is a completely different matter. In Formula 1 anything can happen at any time on track; how could viewers be sure they wouldn't miss key moments in the action due to an ad break? As there was no way of knowing until we had tried it, we were all a little apprehensive.

We didn't have long to find out. Thankfully, Australia went well, Murray was his brilliant best, while Martin was a revelation, a natural talent, adding a completely new dynamic with his energy and level of knowledge to the co-commentator's role. His star was well and truly born. There was some of the expected negative reaction to the advert breaks, but looking back on it we didn't exactly help ourselves. Ordinarily if any sport programme misses something in a break, they replay it when they come back to the action, and the commentator voices it. However, like the BBC in the past, ITV had sold Murray and Martin's commentary on to networks around the English-speaking world, not all of whom took advert breaks at the same time we did, if at all. There was a concern that Murray re-voicing something wouldn't make sense for people in other countries who hadn't missed it, and so it was decided that if we did happen to miss something significant, Jim Rosenthal would come in after the break, and hand over to analyst Simon Taylor who would voice the action. A reasonable idea, but in practice all it did was draw more attention to the fact we'd missed something in a break. It took a year or so for the producers to abandon the protocol and revert to the original idea of getting Murray to recap.

The only thing that didn't go well in Melbourne was what would now be called 'extended workflows'; back then it was known more simply as 'pulling an all-nighter in the edit'. The producers had spent most of the day filming features to fill the pre-race airtime we now had, but had only been able to get started on editing in the late afternoon. I was working on the first race of the season from London, so wasn't on-site to log the tapes and, since the producers hadn't had time to log their own tapes, if they needed some 'paint' shots to cover edits or show examples of what the interviewee was talking about, they had no idea if the two cameramen had shot the required footage, or if they had, where those clips were. Edits, therefore, were taking much longer than they should.

What they needed was someone to log the tapes, someone who knew where every shot was, and on which tape to find it. My job, in fact. Starting from the second race of 1997 in Brazil, it was decided that I would travel to the circuits. To downplay this slightly, the fact was I wouldn't get out much, I wouldn't see any cars on track or be able to leave the TV compound. I was there to manage all the tapes, log them, know where every shot was, log the practice sessions, qualifying and the races and be on hand for as long as was needed. If John 'Noz' Nolan was editing a piece about Ferrari, for example, he would shout through to me, 'Hey, Tedos [his nickname for me], I need a shot of Michael looking happy at the back of the garage, helmet off,' and I'd know that our cameraman Andy Parr had filmed that the day before, or at a previous race. I'd race to the tape shelf, grab the tape and cue the shot up for Noz's approval or rejection. This way, the edits rocked along at a much quicker pace, and not only were there no more all-nighters, but on some nights over that second race weekend in Brazil we actually made it back to our hotel for a Caipirinha and a steak.

Jacques Villeneuve won the Interlagos race, his first victory of the season on the way to the world championship. And on Sunday night, with my first F1 weekend over, I sat in the middle of a block of four seats at the back of a Swissair MD-11 en route from São Paulo to London via Zurich. I couldn't have been happier.

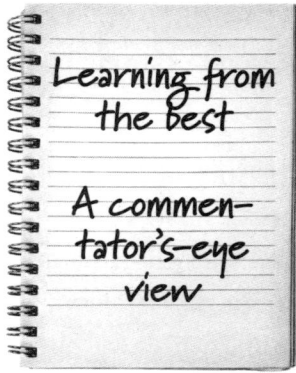

Learning from
the best

A commen-
tator's-eye
view

Chapter 3

The Murray and Martin Show

It is a truth universally acknowledged that no one can go through a career in Formula 1 and be respected by everyone – it's just too competitive an environment for that. However, there was one notable exception, and that was Murray Walker.

Everyone loved Murray, for a whole host of reasons. He was a kind, generous, witty, thoughtful, talented man – qualities which were evident to anyone he met. As a professional, he was widely respected within the F1 paddock, while his irrepressible passion, enthusiasm and joy in the sport communicated itself to viewers at home who responded in kind. Then there was his sheer longevity. Most people who work in F1 enjoy the sport, and probably grew up watching it on TV. If they watched it in the UK, or the USA or Canada, South Africa, Australia, New Zealand or pretty much any other English-speaking country, they would have been listening to Murray Walker's commentary. That inherent familiarity, that people already felt they knew him, meant Murray could walk into any garage or team motorhome or pretty much anywhere else in the F1 paddock and be instantly recognized, welcomed and accommodated in whatever he was looking for.

Most of the time, that was information. Murray always put a great deal of work into his preparation, and he covered a lot of ground on a race weekend. I'd often see him on Thursdays scouring the paddock and media centre, catching up on stories from the week between races, and getting inside information from the teams that he might need to refer to in commentary.

I say 'might' because Murray's style wasn't simply to bombard his audience with facts and figures or stories of an encounter with a driver in a lift. Despite having been at the job for so long, his magic ingredient was that he remained so charmingly enthusiastic about the pure motor-racing aspect of the F1 circus. His enthusiasm was often described as 'childlike', which I think is fair in that, despite having all the answers, he approached the viewing of every Grand Prix from a completely fresh perspective, as a child might have done watching from home. Not letting on that he had a pretty good idea beforehand of what was going to happen in the race, or that he might know the inside story about why a team or driver's performance was set to be the way it was, allowed Murray to connect with his audience, and to share those moments of discovery with them.

That's what made all his emotional reactions so genuine. All of the classic Murrayisms – 'There's nothing wrong with the car, except it's on fire', 'The lead car is unique, except for the one behind it, which is identical' – were instinctive. He said what he was really thinking at the time, and his emotions were never far from the surface. British viewers of a certain age will always remember his lines at the end of the 1996 Japanese Grand Prix, 'He fought from second on the grid, he passed Jacques Villeneuve, he took the lead, he stayed there. And Damon Hill exits the chicane and wins the Japanese Grand Prix – and I've got to stop, because I've got a lump in my throat.' But Murray was never naïve. He knew that his authentic passion and honesty were the best approach, because he understood the power of marketing, selling and promotion from his many years working in the advertising industry (in his former career he had been responsible for advertising Trill budgie seed. His clients had a problem in that their product was so successful, everyone was already using the seed to feed their birds. Murray's solution? 'An only budgie is a lonely budgie.') He knew that in F1 broadcasting, as in any market, you're not selling the steak, you're selling the sizzle, and he understood the psychology behind building tension, conveying excitement and in knowing how get people to feel invested in the outcome. Outside of the drivers I think he did the most to further the popularity and therefore the success of F1.

On screen, Murray never had a cross word for anyone and he served as a moderate foil to some of the strong views voiced by his co-commentator of the day, British ex-F1 champion James Hunt, not known for his discretion ('The problem with Jarier is that he is a French prat, always has been, always will be.') However, off-screen there were some people who were not Murray's cup of tea. He didn't have much time for those who he felt had come into F1 to make a quick buck for themselves while not particularly caring about the sport. Sometimes over dinner the subject of conversation would turn to one of these figures, whether they be brash, arrogant or particularly full of themselves – attributes hardly in short supply in the F1 paddock. Murray would pipe up, 'Oh, him. If he was made of chocolate he'd eat himself,' before breaking out into his huge laugh.

However, he also kept an open mind. Murray was initially suspicious of Flavio Briatore, who had been appointed from the fashion world by the Benetton family to run their F1 team. Murray questioned what someone who had never been to a Grand Prix could possibly know about operating a team. As it turned out, Flavio's work ethic and demonstrable success at Benetton (later Renault) eventually convinced Murray that the Italian's intentions were for the greater good of F1, rather than just himself. Although it was probably a good thing that he wasn't still commentating at the time of the Renault Crash-gate scandal of Singapore 2008 (involving Nelson Piquet Jr and an 'accidental' crash that put his teammate Fernando Alonso in a position to win the race, a crash that later emerged to have been planned). Murray would have thought that 'a very rum do!'

Murray was by no means an apologist for the many occasions F1 didn't exactly show itself in the best light. He would be front and centre in reporting a controversial piece of driving, or a rule breach by a team. He was as disappointed as any F1 fan by an uneventful race, although he'd save his true personal opinion for when we had gone off-air, when he would wrap up the commentary over the podium celebrations, put his microphone down and say, 'Well, that's not going to have them tuning into the highlights, is it?' That would be the extent of his criticism. Murray wouldn't ever call a race boring. Firstly, to him no F1 race could be boring, as viewing these incredible cars going at amazing speeds, driven by heroic drivers, could only ever be 90 minutes of pleasure. Secondly, while there might not be much going on among the frontrunners, there would always be some story or battle further down the field that would hold Murray's interest – and he'd make the rest of us care about it, too.

Alongside Murray there was the reluctant genius: Martin Brundle. Reluctant because he hadn't been ready to start his TV career so soon, and genius because he's the most knowledgeable, most authoritative, quickest-witted communicator in broadcast sport. In his day Martin was a very good racing driver – he was quick, consistent, versatile, mentally tough and smart. When it came to commentary, all that gave him authority, but what he also brought to his television career was an easy charm, a sharp sense of humour and a team ethic worth its weight in gold. Perhaps because he was part of so many good (and bad) F1 outfits over the years, Martin, it turned out, found much to enjoy about being part of a close-knit TV production.

In that first 1997 season the dynamic between Murray and Martin in the commentary box quickly established them both as the stars of our show. I'd been systematically logging their interviews and their recorded links for features. In all the time I spent watching them I was able to learn about how they worked, and about their respective characters. Murray had the habit of beginning his links with a little laugh, to ensure he had a smile on his face when the viewer first saw him, 'making a friendly first impression,' as he put it. Martin wasn't used to this 'TV lark', as he used to call it, and I saw him working away at his takes to achieve something he was happy with.

Up to this point I hadn't really left the TV compound at any of the tracks we visited. I didn't even have a paddock pass, because there wasn't any need for me to go there. I'd arrive at our outside-broadcast truck every morning, log tapes, help out with edits and, when the sessions were on, sit and log those as well. If I was lucky and one of the producers took over logging duties for 10 minutes during a session, I was allowed to lean out the back of the truck and watch the F1 cars speed past.

The view varied. In Monaco we were on the outside of the Swimming Pool exit – a truly breathtaking vantage point from where you could have reached out and touched the cars if you were stupid enough to try. At the Circuit de Catalunya in Barcelona we were on the outside of the last corner, which saw some seriously impressive speeds.

Seeing the cars on track was great fun, and it at least gave me a brief glimpse of the F1 world in which I worked. But after 10 minutes it was back into the cool, dark TV truck for more tape logging. The job that took me to TV compounds in circuits all around the world, requiring hours and hours of my time logging F1 sessions, pit lane camera footage and driver interviews is now done by an AI-powered plug-in to our media library. In 1997 we were a long way off this technological development, but nonetheless halfway through the season there did come a significant change to my job.

Imagine being in the commentary box commentating on the race. You'll need to be keeping an eye on multiple screens showing race positioning and technical data, feeds from cameras following individual cars on track and also watching the parts of the track and pit lane that can be seen from the commentary-box window. You'll be keeping across 20 cars and debating events with your co-commentator as well as listening to information from the production team and your pit lane reporters. What you'd probably wish you had is an extra pair of eyes. And so my boss Neil Duncanson decided to use me as a spotter, helping Murray and Martin filter the huge range of information and ensuring things didn't get missed. Starting at the 1997 German GP, I was sent to the commentary box. If I could find it.

Commentary boxes are interesting little things. At older circuits they're usually located in a position that allows the commentator to see a large portion of the track. Circuits like Silverstone, Hockenheim or Monza run race meetings year-round, and most of these aren't televised live, which means that any commentator present needs to be able to see the action out of the window with a pair of binoculars. Most club-racing circuit commentators only have a timing screen and a good view to work from, and if all the TV pictures go down, that's what F1 commentators need to be able to work from, too.

At Hockenheim the commentary boxes were on a gantry on top of the main grandstand, which meant plotting a route out of the TV compound, which as usual was situated near the paddock on the inside of the circuit. I had to go through a tunnel under the track, come out behind the main start/finish line grandstand, find the security guard by the gate, and then climb up four storeys to the top.

Once there, my instructions were no more detailed than simply to 'give Murray and Martin a hand'. I soon realized that having an extra pair of eyes on the timing screens and somebody to act as a halfway house between producer and commentator was going to be a big help. Not least because, while Murray could commentate accurately and entertainingly for the duration of a Grand Prix, the moment somebody talked to him through his headset he would grind to a halt, looking around the box as he listened. In TV terms, Murray couldn't take talkback.

That's not entirely fair because, as a TV professional, Murray *could* listen to a director's voice in his ear when he was conducting an interview, or leading to the next part of a programme. However, if he was interrupted when he was concentrating and commentating in full flow he would usually stop talking and look at me as if to say, 'Who was that, what were they saying, and why the blazes were they interrupting me?' All important thoughts, but with no words emerging at the same time it did leave a bit of a gap in the commentary.

I had noticed that he was fine with talkback if it was simply telling him to start ('Cue Murray') or to stop ('Wrap it up please Murray'), but anything outside of that was a problem. I wondered if a series of handwritten signs might be a more effective alternative.

I got through 50 sheets of A4 paper during that first weekend in Hockenheim. On my return to London I decided to print out the five most common instructions in big, bold, capital letters, and get them laminated for regular use. Those five cue cards were:

START: An obvious one, I thought. I could have put 'GO', but Murray might have bolted for the door!

LEAD TO BREAK: This was a difficult one, as we tried to make the throw to in-race advert breaks as subtle as possible, and if I'd put 'GO TO BREAK', Murray might have said, 'So, it's lap 35 and we're going to a break', whereas 'lead' reminded him to gently wrap up his sentence or thought and take a pause in order for the production gallery to roll in the advert break.

JAMES IN PITS: Our pit reporter James Allen was always in the pits, but this sign had a dual meaning. If I held it still, it meant James's report wasn't urgent or particularly time sensitive, whereas if I started to wave the sign up and down, that meant something urgent was happening in the pit lane, and Murray or Martin should try to throw down to James as soon as they could. I had what's called a 'snoop' of James Allen's and Louise Goodman's microphones through my headphones, so I could hear everything they said to the producer. Very useful, as it helped me gauge whether something was urgent or not, and very useful training for my next job!

THROW TO LOUISE: Unlike James, Louise wasn't always in the pits. Often, she was with a driver in the paddock or a team boss in a garage, so we felt a more generic 'throw' was best.

WRAP UP COMMENTARY: I'd considered 'STOP', but feared, Murray being Murray, he would just stop talking immediately. I'd thought long and hard about 'NO MORE COMMENTARY', but that sounded a little curt, so 'WRAP UP COMMENTARY' won the day.

The cue cards worked perfectly. Most of the time they would be greeted by a thumbs-up 'received and understood' acknowledgement from Murray, but sometimes he was so engrossed in his commentary that he would either not notice or ignore the cards. At which point I'd start to move them gently up and down in his field of view, much to Martin's amusement.

That first weekend in the Hockenheim commentary box, with the huge swathes of Schumacher fans below us in the grandstand enjoying the German GP, was a markedly different experience to logging the race in a windowless broadcast truck. I was standing alongside Murray Walker, my TV hero, the man whose every word

I'd hung on to every other weekend for years, and Martin Brundle, the new star of F1 broadcasting. And whether it was spotting a spin, or a quick sector time on the timing screen, or giving Murray a steer on which driver the visual feed had cut to if he'd been looking elsewhere, I was helping. As you might well imagine, the four years I spent in the commentary box remain the most enjoyable time I've had in my professional life.

As the season continued, I'd spend more time in the box on Fridays, setting up the TVs and fixing sun-screening material to reduce the light glare from the windows. Murray would watch every practice session from the box (live coverage of free practice only started in 2012), and he used the time to prepare for qualifying and the race. Martin might join us for the second or third practice sessions, but would generally alternate between watching the action trackside, from the media centre, or from a motorhome over coffee with a team member. For Murray, practice sessions were a vital part of his preparation, research and thinking time.

He would get to the commentary box around half an hour before first practice and get it set it up just the way he liked it. It was identical at every race. His preferred commentary-box layout was simple. He'd always be on the far left of the box with his notes and statistics stuck to the left-hand wall or window. Directly in front of him was the main F1 world feed (the stream of images produced by F1 and fed out to broadcasters around the world) on a high-quality TV placed on a packing case, so it was just below his eye level. The timing screen was then put on top of the main monitor, and in front of that we placed a solid plastic tripod case that came up to Murray's midriff. This acted as a very handy little writing table on which Murray would take note of who pitted on what lap.

In the middle of the commentary box was a monitor showing the ITV programme output, so Murray and Martin could watch our pre- and post-race coverage. Martin had his own world feed and timing screens centre right, and squeezed into the far-right corner was my little spot with a talkback panel to communicate with the producer and director in the gallery, volume knobs to listen to James and Louise in the pits, and switches to talk via the headsets to Murray and Martin.

Once everything was set up to Murray's satisfaction he would sit and watch the practice sessions with an intensity that took me a while to understand before I realized that he was practising too. He was watching the cars, making sure he was spot-on with his identification, and noting features of the circuit itself. He had a copy of the track map, and he would annotate it, marking which advertising billboards were on which corner. This was important, because if he saw a car stop, or if there was an accident, his first job would be to identify the car and driver. The second job would be to identify the corner, and if for example there was Foster's signage in the background then Murray knew from his sponsor marked-up map that the incident was at Turn 5, or if there was a Rolex billboard, it had to be Turn 10, and so on. When first practice ended, Murray would take his headphones off, slap his thigh and say, 'Right! That's lunch,' and we would both make our way back to the TV compound.

I have been lucky enough to parallel Martin on his second career journey through broadcasting. We first got to know each other at ITV Sport, and then moved together to the BBC, and on to Sky Sports. In another life, I would love to have been his race engineer, but in this one we started out having to communicate by looks and gestures alone, as we stood shoulder to shoulder in the commentary box. From that point, Martin and I developed a kind of unspoken communication system over the years – we can catch each other's eye for a second across a grid or paddock and know exactly what we're each thinking. His nickname in the TV compound is 'The Guv'nor'. I don't know why, it just fits. Probably because he is the best in the world at what he does.

But back to my early commentary-box days and I was concentrating on everything so hard I sometimes had to remind myself to enjoy the little behind-the-scenes moments. One year at Japan's Suzuka circuit Martin came up to watch the second free practice session with Murray. It wasn't a particularly eventful session, so we all sat there in silence. After 10 minutes or so Martin shared an observation with Murray about a particular racing line a driver was taking. 'Oh yeah,' said Murray, before they both lapsed into silence for another 15 minutes. We were all battling the effects of a plus nine-hour time difference, and the soporific engine noise made it difficult for Martin and me to beat the jet lag and stay awake. Murray, though, was glued to that TV. Another 15 minutes passed and just as our eyelids were starting to droop, the race feed cut to a car spinning off, fast, at the 130R corner. 'WHAM!' exclaimed Murray at full volume, shattering our drowsy silence. 'Oh, crikey!' A pause to check the driver was conscious. 'Dear oh dear, he's made a right mess of that one, hasn't he?'

Martin nodded in agreement as the Japanese marshals ran to help the unfortunate driver who had flung his Grand Prix car into the gravel. Murray made a note of who and when. Another couple of minutes into the recovery process passed before Martin, lifting off one side of his headphones, had something to say.

Martin: 'Charlie's got a Bobcat.'
Murray: 'What?'
Martin: 'Look, Charlie's got a Bobcat.'
Murray: 'Oh, yeah.'

And with that, they both put their headphones back on, stared back at the screen, and settled into another 20 minutes of silence.

I smiled. Only in F1 could the words 'Charlie's got a Bobcat' make any sense. Martin was pointing out that FIA (Fédération Internationale de l'Automobile) race director Charlie Whiting had deployed an additional recovery vehicle, a small crane made by the Bobcat company. Martin hadn't seen Whiting sanction the use of a Bobcat before, and clearly thought it was of note. Murray was less interested but I loved that little exchange, and I'll never forget it. Charlie's got a Bobcat. Great name for a rock band.

The next day, it was showtime. Murray had a very specific warm-up routine. With about five minutes to go before the F1 introduction that signalled it was nearly time to start commentating, Murray would stand up and take off his headphones. This was my cue to remove his chair to the back of the commentary box or place it outside. He wouldn't be needing it again. Then he would start his exercises. He proceeded to stomp his feet three or four times left to right, and then began a vigorous upper-body aerobic warm-up session. Arms out to the side, five times. Alternate forward punches, five times. Elbows back to stretch and open up the rib cage. A few more tightened fist pumps and he was ready for action.

Murray's commentary was a physical performance. As he commentated from a standing position he felt a warm-up opened up his chest, which delivered a better, more powerful vocal performance. After that it was time for the mouth exercises. The story goes that a famous theatre actor (I recall Sir John Gielgud being mentioned) once told Murray that the mouth contained a hundred muscles, all of which need warming up if one was to deliver one's best vocal performance. Murray would duly go through the most rigorous contortions of his mouth to get all those muscles well and truly warmed up. I must admit, it works, and if your job depends on you using your voice it's worth noting. Most of the time our mouths are shut, and we talk from our lips. To really enunciate fully and not trip over your words, you have to warm up your mouth. It's a practice I learned from Murray, and I make sure to do it before every session.

A few more shoulder rolls and neck stretches, and then the F1 titles played, signalling five minutes until the race start. 'Cue Murray!' and we were off. I never knew how Murray was going to begin the commentary, but what you could expect was that it would be dramatic. *'Stand By Because This Is The Big One!'* Murray might have said that at the first race of the season, or at Silverstone, or at the world championship decider – to him they were all big ones! He would set the scene, talk about what had been happening leading up to the race and recapping qualifying for those who hadn't seen it. All of this allowed Martin time to come up to the box after his live grid walk. Once he was settled Murray would bring his co-commentator into the conversation for the first time.

As Murray might have said, with 'tension so thick you could cut it with a cricket stump', the cars lined up on the final starting grid, and it was 'go' time. The FIA only started using the current five lights on-hold-and-off system at the beginning of 1996, so Murray used to mix up his 'lights out' phrases. He was most well-known for 'And it's

Go, Go, Go,' which worked well with his inimitable voice. Honestly, it didn't really matter what he said, all Murray needed was to buy a few seconds to allow the cars to move off the line so that he could establish whatever order they were in. The fact that he used to mix it up always felt, to me at least, that he was having fun, and wanted the flexibility to call the start of the race in whatever way he liked.

And there you have the key aspect of Murray Walker's commentary style – he liked what he was watching. That description of Murray's enthusiasm having a 'childlike quality' is true in part and goes some way to explaining the connection he made with young viewers who got hooked on F1. I would add that he simply had a genuine passion for motor racing and thoroughly enjoyed what he was doing. His responses to events were unrehearsed: 'Oh my goodness, look at that!,' or 'This is the worst start to a Grand Prix that I have ever seen *In The Whole Of My Life.*' Murray's use of language was always exhilarating.

Giving yourself time to think when you're talking consistently for hours on end is an art, and Murray managed to achieve this by slowing down and carefully managing all the available content. If you listen back to any of his race commentaries you'll notice that, like one of his beloved BMWs, he had his own personal six-speed gearbox. He would start in fourth, go into fifth for the start, before dropping down through the gears as the race progressed. He'd be able to ratchet back up to fifth and even sixth gear if the action was particularly significant, but knew very well that the audience would find a commentary constantly in sixth gear completely exhausting to listen to, so he used that mode sparingly.

Gearbox perfectly tuned, a certain economy with words allowed him to think about what to say next. Purring along in second or third gear, Murray was careful never to say too much, or to give the listener an overwhelming amount of information. 'So, lap 45 in Hungary. Ayrton Senna leads for McLaren with Nigel Mansell 2.3 seconds behind. Gerhard Berger is third, a further 5.8 seconds behind the Williams Renault.' A typical succinct Murray recap with information gleaned from the timing screen, but it tallied with what the viewers could see on track, and reassured the audience that Murray was on top of the whole situation.

When he'd finished a top six rundown Murray used to physically push his microphone away from his mouth, a signal for Martin to pick up. If there wasn't much overtaking or racing to get excited about, that was when Murray would turn to the stories that he'd researched on Thursday. Choosing three or four topics of conversation in the paddock that weekend, Murray would collate and prioritize which stories to share, for instance the latest moves in the driver market, or if a team principal or chief designer was being appointed or replaced. These topics were usually delivered in first or second gear.

Finally, there was that voice. Not before or since has there been a voice that was so naturally suited to the sport it covered. It was powerful, it had immense physical volume, and it had a recognizable and repeatable tone. Generations of viewers can, to this day, immediately conjure up Murray Walker's voice. To listen to Murray in top gear was the next best thing to being trackside yourself. Through Murray you knew that what you were watching was exciting, significant and important, that it meant as much to him as it did to you, and that is why he was so loved by his audience.

Chapter 4

Jacques, Gilles and Jerez

———

Not only was I now travelling with the Grand Prix circus and in a position to watch the races from the commentary box, but my first season ended with one of the all-time great stories of F1. Before Lewis Hamilton was out of karting and Max Verstappen was out of nappies, this was the season when Michael Schumacher failed to pull off the trick he'd been successful with three years previously: running into the side of a Williams to try to win the world championship.

In 1994, at the Australian Grand Prix, the Williams in question had been Damon Hill's, its front suspension damaged beyond repair following a sideswipe from Schumacher's Benetton. This time it was Jacques Villeneuve's Williams that was hit by Schumacher's Ferrari as the Canadian was trying to overtake for the lead of the European Grand Prix. I watched the whole thing unfold from the commentary box at the Jerez Circuit in southern Spain, with Murray Walker, trousers metaphorically on fire (a description first coined by Clive James that more perfectly sums up

Murray's approach to race commentary than anything I've read before or since) and Martin Brundle, whose cool racer's reaction precisely summed up the desperate manoeuvre: 'That didn't work, Michael, you hit the wrong part of him, my friend.'

The cars may have collided in that championship-deciding race, but the story of the Villeneuve–Schumacher clash began at the start of the 1996 season when they found they shared a common adversary in Hill, Schumacher's victim from Adelaide '94, now Villeneuve's teammate. Jacques came in hot to the Williams Renault team from IndyCar in the USA, where he had won both the 1995 title and the famed Indianapolis 500. He was at the top of his game in the American racing scene, but the lure of Formula 1 had been irresistible, not least because of his F1 heritage.

Jacques's father was Gilles Villeneuve, French-Canadian but European in spirit and adored by F1 fans for his flamboyant, acrobatic driving style, his sheer speed, his intense competitive spirit and his affection for Enzo Ferrari, who, in turn, loved Gilles like a son. Tragically, Gilles was killed after an accident in his Ferrari during the qualifying session for the Belgian Grand Prix of 1982. Jacques was just 11 years old and grew up with the double burden of missing his father, but also, when he himself started racing, with the weight of expectation that came with being the son of the great Gilles Villeneuve. Added to this was the pressure he put on himself, wanting to win the F1 championship his father never did.

The Adrian Newey-designed Williams FW18 was quick, and a winter of intense testing saw Villeneuve extremely well prepared for his debut season, so much so that he led the first race of 1996 in Melbourne from pole position before an oil leak forced him to slow down to preserve his engine, handing Hill the win. Villeneuve gave Hill a good run for his money that season, making the top step of the podium four times, but Damon's eight wins saw him eventually crowned champion at the end of the year in Japan.

Was it the seeming ease with which Villeneuve was able to match Hill's pace that led Frank Williams to lose faith in Damon? Although Frank remained convinced of the British driver's race craft and consistency, he decided, in a move that surprised everyone, that he saw more natural ability in Heinz-Harald Frentzen, a funeral director's son from Mönchengladbach who had matched Michael Schumacher lap for lap when they competed in German Formula 3 (in the sort of detail that delights me you might be interested to know that Frentzen returned to the undertaking business after his retirement from Formula 1). In 1997, with Frentzen replacing Damon Hill at Williams, Villeneuve had a new set of variables affecting his own championship challenge.

Things started off well enough. Jacques won two out of the first three races, but Frentzen was on the pace and pushing hard. Meanwhile, McLaren's (first) partnership with Mercedes engines was starting to bear fruit and the car was quick, if not particularly reliable. As the battle with Schumacher ebbed and flowed, Villeneuve won his seventh race of the season at the Nürburgring on a chilly late September day. It would turn out to be his final race win in Formula 1, the beneficiary of an embarrassing double-engine failure for McLaren duo Mika Häkkinen and David Coulthard on Mercedes's home turf.

I loved the Nürburgring. The F1 circuit was well laid-out, the expert organization highly efficient and we would drive past the final straight of the old circuit, the famed Nordschleife, in our rental cars every morning. We stayed in a town called Blankenheim, which had the Eifel mountains on one side and the 'Schloss Blankenheim', an imposing medieval castle, on the other. Should you be curious to experience it for yourself, you can stay there – it's now a youth hostel.

After the Nürburgring there were only two Grands Prix left in the 1997 season and Villeneuve well and truly drew a blank at Suzuka. Already compromised by a suspended race ban imposed for failing to slow down sufficiently in an accident zone at the Italian GP, Jacques repeated the offence and was disqualified from the Japanese GP – a penalty so harsh as to be unheard of these days. For Jacques it meant that having gone into the Suzuka weekend nine points ahead of Michael, he started the final round in Jerez one point behind.

In previous years the season had ended at an Asian or Australasian race, but this year we finished in Europe, giving the F1 teams an opportunity to achieve the maximum marketing impact in their home territories, and none were keener than McLaren, who had launched their car eight months previously at the Royal Albert Hall with an exclusive performance by one of the biggest musical acts of the nineties, The Spice Girls. McLaren were on the verge of something big, competitively speaking. Having left Williams for McLaren at the end of 1996, Adrian Newey had designed a car that was now reaching full potency and Mercedes were getting on top of the small reliability issues that had previously been their undoing. McLaren boss Ron Dennis and the amiable Norbert Haug, motorsport director of Mercedes, were Formula 1's power couple. In sharp contrast to the 'tyre-marks on the garage floor', 'oily rag' aesthetic of Williams, McLaren was positioning itself more like a science-lab-style F1 operation, a paragon of excellence and efficiency, the pinnacle of Formula 1 technology.

Selling that idea was down to good old-fashioned marketing and McLaren had its own well-financed marketing division headed up by Ekrem Sami, a legend in F1's commercial world. On the eve of the Jerez weekend, McLaren Mercedes held a lavish party for their team and sponsors at one of the town's sherry wineries. The travelling F1 media were sent an invite which read 'Take the Bull by the Horns'. Nobody really knew what this meant: some assumed it alluded to McLaren's determination to end the season on a high, while others saw it as a reference to Spain's bullfighting tradition. It was only years later I found out it pointed towards a design detail of the McLaren's car livery that had remained practically unnoticed for years: in the negative space in the middle of the car's front wing was a silhouette of a bull's head and horns. The party was a great success, enlivened by McLaren's team co-ordinator Jo Ramirez, a proud Mexican who lived in southern Spain, holding court with stories of the explosive battles between Ayrton Senna and Alain Prost of F1 seasons past.

From a memorable party to a memorable qualifying session. For the first time in Formula 1 history, three drivers set identical lap times in qualifying, a 1.21.072. Murray Walker called it 'almost unbelievable' and there were plenty of cynics in the F1 world who spoke of timing glitches or some kind of manipulation designed to increase the tension ahead of the race. Williams later published the telemetry data proving definitively that Jacques Villeneuve and Heinz-Harald Frentzen had completed their laps with exactly the same lap time, to the thousandth of a second. Ferrari didn't release the data from Michael Schumacher's car but in the end, it didn't matter – Villeneuve had set the time first, followed by Schumacher and then Frentzen, so that was the grid.

As for the race, while it ended for Michael Schumacher in that controversial block on Jacques Villeneuve, that was by no means the final twist. After it had been struck by Schumacher's front-right tyre, Williams were concerned about the integrity of Villeneuve's car. Rightly, as it turned out – post-race inspections revealed that the on-board battery keeping Villeneuve's electronic systems alive had become dislodged by the impact of the Ferrari and was left hanging by its red-and-black connection wires inside the car. The thought of this battery, weighing half a kilo or so, dangling freely in the Williams's left-hand sidepod, secured only by a couple of wires, must have given Jacques chills for years after. Driving to achieve maximum speed while protecting his car as he did meant the battery survived the race and he emerged the world champion. Had he not done that, Michael would have won Ferrari's first drivers' title since 1979.

As all F1 viewers of that season cannot fail to have been aware, Michael had some assistance in the form of a couple of friendly drivers. His Ferrari teammate Eddie Irvine irritated Villeneuve by holding him up in practice several times. Springing vividly to mind is the moment Jacques hopped out of his car at the end of Friday practice, strode down the pit lane and confronted Irvine, who was still in his Ferrari, being wheeled back into the garage. It's hard to convey anger when your face is hidden by a crash helmet, but no viewer could be in any doubt that Villeneuve was furious and that Irvine got the message. Years later, Jacques claimed his was an intentional overreaction, as he wanted to draw attention to what he saw as Ferrari's 'games'.

In the race, too, Ferrari had allies. Halfway through, Villeneuve lost time as he was held up behind Sauber driver Norberto Fontana, which led to this exchange between Murray Walker and Martin Brundle in the commentary box that made me smile:

Murray: 'Case of champagne from Ferrari to Sauber for Norberto Fontana because the Argentine newcomer really, really helped Michael Schumacher on his way there.'
Martin: 'What engine have they got in that Sauber, Murray?'
Murray: 'Er . . .'
Martin: 'Isn't it a Ferrari?'
Murray: 'Well, it is, yes. Oh Martin, you are a cynical chap!'

Following the lap 48 clash with Schumacher, Villeneuve led the race until the last lap. Villeneuve then allowed both McLarens through, giving Mika Häkkinen an emotional maiden win just two years after a qualifying accident in Australia that had nearly cost the Finn his life. Afterwards Villeneuve explained that he had needed to protect his car following the clash, and that he wasn't bothered about finishing on the podium: he had the points he needed to be world champion. There were, however, two key events that prompted the FIA to briefly investigate McLaren and Williams for collusion or 'race-fixing' once the season was over.

First was the appearance of Williams technical director Patrick Head in the McLaren garage, a few laps before the end of the race. James Allen had positioned himself to watch the race from the same garage and ITV cameraman Keith Wilson was also on hand

and filmed the ensuing chat. Ron Dennis shielded his mouth from the camera. Patrick Head was known to be hard of hearing, but Ron was also in the habit of covering his mouth to shield his words when he spoke, and this wasn't by any means the first or the last time I observed him doing this.

It was a highly unusual thing to happen and naturally viewers suspected that some kind of deal for Villeneuve to let the McLaren duo through was being confirmed. Upon investigation, however, the FIA found there was no case to answer and Williams simply called it 'good communication'. McLaren denied there had ever been a pact and maintained that in fact they had simply been honouring an informal understanding not to get involved in the championship decider. This claim was then supported by the second event: a team radio message, that came to light some time after the race, which revealed engineer Jock Clear advising Villeneuve of the closing McLarens. On tape you can hear Clear asking Villeneuve to 'remember what we discussed' and reiterating, several times, how Häkkinen had been 'helpful' to the Williams cause.[1]

For Mika Häkkinen the race victory was the start of a winning streak that would lead him to his first world championship the

---

1   Jock Clear (to Villeneuve): 'Keep concentrating, Jacques. Keep concentrating. Häkkinen up to position two. Häkkinen quite quick and very helpful. (Later) Be aware that Häkkinen is now in position two. He probably wants to win. Very helpful. (Later) DC [David Coulthard] is controlling [Eddie] Irvine. Häkkinen immediately behind you, Jacques. Immediately behind you, Häkkinen.'

Clear (Later): 'Keep concentrating, Jacques. Häkkinen is immediately behind. Last lap. Last lap. Häkkinen has been very helpful. Jacques, position two. Don't let me down, Jacques. We discussed this . . .'

following year. For Villeneuve, meanwhile, Jerez was all about the release of a season's-worth of pressure and an opportunity to celebrate achieving his long-held dream of winning an F1 Championship. On his return to the pits, Villeneuve hugged Clear, whose own ambition to beat Schumacher (a driver he would later work with closely at Mercedes) had also been well and truly achieved. 'We did him! We fucking did him!' screamed Jock, as the Williams mechanics, sporting yellow-blonde wigs to mimic Jacques's peroxide hairstyle, carried the Canadian on their shoulders down to the podium and a moment that, although brief, would remain bright in my memory and that of millions of other F1 viewers worldwide. Jacques was now a champion in his own right, no longer just Gilles's son.

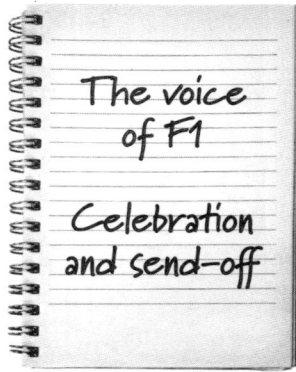

Chapter 5

Murray's Last Season

The Belgian GP at Spa-Francorchamps has always struggled with traffic problems. It's not the organizers' fault, it's just that there's a huge number of spectators to get in and out. The race is enormously popular among Dutch, German, British and French fans. In fact, the only country that doesn't appear to be well represented at the Belgian GP is Belgium! It's a circuit with one road in and one road out. Sometimes, if the police are in a good mood and if there's a 'T' in the month, they might open some of the single-track lanes that lead out of the back of the huge five-mile circuit, allowing better dispersal to all points south and east. But most of the time everyone is funnelled into the same massive traffic jam northbound towards the motorway. And it was after the 2000 race that I found myself in this very traffic jam, occupying the back of a hire car with Murray Walker, on the way to Brussels Airport.

Louise Goodman was driving with James Allen alongside and Murray was next to me in the back. It was rare for Murray to travel in our reporters' car – usually he'd get a lift with Jim

Rosenthal, Tony Jardine or ITV's technical producer, Roger Philcox, a legend on the production crew. Roger's job was to book all the satellite lines for ITV's F1 programmes, and he used to travel around the world with a briefcase stuffed full of US dollars in case he needed to pay for an extension to a satellite booking at short notice. Back then there wasn't the ability to transfer money instantly at the click of an app, so in the event of a race overrun or a red-flag stoppage, Roger would save ITV from falling off-air by running to the uplink van and paying in person for the satellite fees with wads of the network's cash.

Unusually for Spa-Francorchamps, the 2000 race hadn't overrun, which had enabled Roger, Jim and Tony to escape eastbound to Cologne Bonn Airport across the border in Germany, from where they had a flight back to Heathrow. (From Spa, it's actually easier to get to Cologne than Brussels, despite it being in a different country.) In fact, due to his many frequent flights keeping ITV's outside broadcasts on-air worldwide, Roger Philcox held such high status with British Airways that he was able to phone a special number to request that the plane he was booked on not leave without him – thus making travelling alongside him extremely advantageous. To my knowledge Roger never missed a flight.

Murray preferred a more relaxed kind of race weekend getaway, often electing to stay Sunday night near the circuit or in an airport hotel to rest after the day's exertions, rather than rushing to get home that evening with all the stress that entailed. He also factored in the fact that if his flight was to arrive back at London Heathrow, that meant there was still a 90-minute drive south to his home in the New Forest. Best make that journey when you're not absolutely knackered, Murray reasoned. In this case, he was overnighting at a Brussels Airport hotel, so it mattered not one iota to him that we were moving at a snail's pace out towards the motorway.

Murray was a fantastic travelling companion, because he loved to pass comment on everything – on the race that he'd just commentated on, or the particular performance of a certain team or driver, or sometimes the standard of road driving in the queue around us. Usually, it was in his conversational voice, but sometimes he'd amuse his fellow passengers with a burst of 'and it's go, go, go!' when the traffic lights turned green.

In this instance we had been turned away from our usual route towards Liège, and were being vectored into an unfamiliar direction. In the front, Louise and James were poring over the map we'd been given by the hire-car company and trying to figure out a new route (younger readers, this is what we did before Google Maps and SatNav). While the driver and her navigator were getting more and more frustrated in the front, I noticed Murray gazing out of his window. He then theatrically leaned across me to look out of my window. Louise and James had given up complaining about the diversion and settled into a 'We'll get there eventually' mode. Only then, with perfect timing, did Murray lean forward and say, 'Do you know, I think I drove my tank through here in 1944 . . .'

We laughed, as it put our minor squabbles about the vagaries of Belgian traffic into perspective. Here was a man who at the age of 76 was not only still very much at the top of his game as the voice of F1, but in his younger days had served in World War Two as an officer in the Royal Scots Greys regiment.

Murray's age was never an issue as far as the ITV production was concerned. He was consistently brilliant at what he did, and was adored the world over. Sure, there were little mistakes that he made, the so-called 'Murray-isms', but as I'd find out in my own time on-air, we all make mistakes, it's just when you make them on television, everyone knows about it!

For a commentator, accurate driver identification is the most important thing to perfect, even though the commentator has a much less favourable viewing environment with sunlight coming in from all angles of the average commentary box, and only, in those days, a small 4:3 cathode-ray tube TV to watch, amid constant requests on talkback providing multiple distractions. The viewer, sitting in comfort at home watching on a massive TV, is always going to have an easier time identifying cars than those in the broadcast booth. That's why, as Murray's commentary-box producer, I tried hard to limit the distractions, leaving him free to concentrate on his world feed monitor and the timing screen on top of it. If Murray did mis-identify a driver, I'd wait for Martin to gently put him right, or if it was important, I'd press the little talkback button I had going into his headphones, and just say the name of the driver. Murray would then correctly identify whoever we were looking at, and we'd all move on. It really wasn't a big problem.

Having said that, Murray cared very much about maintaining his own professional high standards. He'd been doing the job for over 50 years. Despite having had two hip replacements, he was physically active and as fit as a fiddle. The only things he did suffer from health-wise were the occasional coughs and colds he picked up because of all the air travel. Long before any of us started cleaning surfaces with antibacterial wipes, Murray had noticed the cause and effect between long-haul flights and minor illnesses, which I remember he found annoying.

It was a happy time to be a part of the ITV F1 team. By the end of 2000 we'd had three different world champions, and we were just entering the Michael Schumacher era, when the great man would win five straight drivers' championships and return Ferrari to the top of the sport. Viewing figures were good, and people even seemed to be getting used to the mid-race advert breaks. Thus, when a rather nasty column about Murray appeared in the British tabloid the *Daily Mail*, none of us on the production team paid much attention to it. It was followed up the next day by a readers' poll on whether it was 'time for Murray Walker to hang up the microphone'. It wasn't, of course, but Murray used to buy the *Daily Mail*, and perhaps because it was a newspaper he read, it wasn't so easy for him to dismiss it as just 'next day's chip wrapping'.

In his autobiography *Unless I'm Very Much Mistaken*, Murray later revealed that, being 'a sensitive soul', the article had seemed like a warning shot across his bows. Annoyed with himself for a mistake in which he'd mixed up Michael Schumacher and Rubens Barrichello at the German GP a few weeks earlier, Murray had begun seriously to think about whether it was better to stop when he felt he 'was still at the top of the tree rather than tumbling down it'. With retirement in mind, Murray went to see ITV's controller of sport at the time, Brian Barwick, a genial Liverpudlian who later became more widely known as the chief executive of the Football Association. Barwick told Murray that it had to be his decision, and that the network would support him either way. In hindsight, I believe Murray went to see Brian hoping he'd be talked out of the idea of retiring, but if that was the case he got no such response. If they're not going to beg me to stay, Murray may have reasoned, maybe it is the right time to go.

The outcome of that meeting was the mutual decision that Murray would be given a year's farewell tour, and if 2001 was to be his final season behind the microphone, it would be a celebration. Murray also agreed to miss three races out of the 16 that year, essentially so that ITV could field test his successor, none other than current pit lane reporter James Allen. And, significantly, as it turned out for me, they also planned to try out a couple of candidates for James's old job.

Over the winter, Murray's retirement became public knowledge. Within the F1 community and particularly, of course, within our ITV production, there was sadness at the end of an era, but also determination to send Murray off in style. Murray's swansong year kicked off memorably at the first race of the season in Melbourne, where the organizers went to town for him. Murray was hugely popular in Australia, and he returned the nation's regard. He would usually fly out there a fortnight before the race with his wife Elizabeth, and they would tour around a different part of the country or spend a week on a cruise before making their way to Melbourne. On Sunday morning the grandstands of the 2001 Australian Grand Prix were packed with well-wishers as the drivers emerged for the routine track parade.

Murray had been doing some research in the pits when Stuart Sykes from the Australian Grand Prix Corporation tapped him on the shoulder and suggested he might like to go out on to the grid. Intrigued, Murray crossed the pit lane to discover that one of the vintage cars had been stickered up with 'Murray Walker' across the windscreen sun strip. A gobsmacked and delighted Murray was driven round at the back of the parade and received as big a cheer as any of the drivers.

The commemorations and special events continued throughout the season. One of Murray's goodbye presents was a ride in the McLaren two-seater, a racing car well ahead of its time – it even had its own chassis designation, the MP4/98T. McLaren boss Ron Dennis and his marketing maestro Ekrem Sami had invested a lot of time and money into the two-seater project. They recognized that it was a huge PR opportunity for McLaren to be able to put their sponsors and VIP guests in a car in which they could experience the thrill of Formula 1, but with state-of-the-art safety standards. The two-seater was well engineered, well built (with a race-grade Mercedes V10 engine) and was driven most often by David Coulthard, Mika Häkkinen or Martin Brundle.

Murray's outing in the MP4/98T took place at Silverstone, a few weeks ahead of the British GP. ITV had a crash helmet specially designed for him, and it was my job to go and collect it from a company called Grand Prix Racewear. It had Murray's name on the side and the Union Flag on the top. Martin Brundle was going to drive Murray, and if I'm honest, there had been some qualms about putting this elderly legend in the back of an F1 two-seater. What would happen if he passed out due to the G-forces? What if something broke and they crashed? Murray had driven a McLaren himself back in 1981, but that had been years before, and maybe his body wouldn't be up to it now. As a precaution there was a panic button, which if pressed would alert Martin to slow down and return to the pits. Need we have worried? Of course not. Murray had the time of his life; he absolutely loved it. Brundle didn't hold back, and really gave his passenger a flat-out lap. The edited film aired over the build-up to the British GP, and you can still watch it online. The look of delight on Murray's face afterwards was priceless. 'Well, I can tell you this,' he said once he'd got his breath back. 'If you ever think that life is dull and ordinary, and that things are passing you by, you should try this.'

We wanted to make Murray's last home Grand Prix as special as possible, but with so much having been done already we found ourselves slightly at a loss as to what that something special could be. From getting Murray a ride with the Royal Air Force Red Arrows to lowering him out of a Royal Navy Sea King helicopter on an abseiling rope to land at the front of the starting grid, some seriously wild ideas were being considered.

At this point one of the producers asked me to run to the tape library and pull out a VHS recording of the previous year's race so we could assess how big the area was at the front of the grid for a potentially abseiling Murray Walker to land. Watching the tape we noticed something. The British fans loved to bring their own handmade banners and flags supporting their favourite teams and drivers to Silverstone, hanging them on the fences or the grandstands around the track. Why not ask the fans to bring their own flags and banners with messages written on them for Murray? Jim Rosenthal put out an on-air call for F1 fans to get involved. And the Silverstone public really did rise to the occasion. Come race weekend the grandstands were lined with 'Murray Walker' flags and banners with amusing Murray quotes. As the host broadcaster at the British GP (before F1 standardized and centralized the main race feed coverage) we were able to make the most of it by choosing the perfect moments to cut to these banners on the coverage when Murray was watching from the commentary box. He found the messages from the fans very touching and was full of emotion when, on race day, the FIA and Bernie Ecclestone sent him out on the drivers' parade once again.

But Murray's most raucous send-off during that 2001 season came appropriately enough from Paul Stoddart, the gregarious Australian entrepreneur, then owner of the Minardi team. 'Stoddy', as he was known, was a firm friend of Murray's. Alongside his F1 interests he ran European Aviation, a charter and cargo airline based out of Bournemouth Airport. As a side note for anyone as interested in the world of aviation as I am, the airline exists very successfully to this day. It's now known as European Cargo, and operates a fleet of Airbus A340-600s, massive four-engined aircraft bought at good prices from Virgin Atlantic. In 2020, Stoddart and European Cargo made a significant contribution to the UK's Covid-19 pandemic response, ferrying testing kits and protective equipment from China to the UK when they were needed most.

As Murray's house in the New Forest was only 20 minutes from Bournemouth Airport, Murray was a regular on Paul Stoddart's passenger aircraft, a BAC 1-11 that first entered service with British European Airways in 1969. From Bournemouth, the aircraft would hop up to Coventry to collect mechanics and engineers from various teams before flying on to whatever European race was being held that weekend. This arrangement saved Murray hours of travelling time.

Monza was the last European F1 race for Murray, so for the home-bound leg after the race on Sunday, Stoddart pulled out all the stops to make it a flight to remember. He decorated the side of his BAC 1-11 with a giant decal reading 'Goodbye Murray and thanks for the memories'. He had also invited a few more people to come along for the celebratory flight in addition to the returning crowd of mechanics and engineers. To my eternal gratitude my producer Rupert Bush asked myself and Andy Parr, our top cameraman at ITV F1, to go along to document the occasion. Throughout the weekend we all had to keep the planned celebration a secret from Murray. We left the circuit before him and joined the other passengers to line up on the tarmac at Bergamo Airport wearing special 'Thanks for the Memories' T-shirts. I still have mine. When Murray arrived, he was delivered straight to the side of the plane and found all of us in a guard of honour waiting for him. I think he may have suspected something was up when he was chauffeur-driven to the door of the aircraft, but the sight of all of us clapping him in our T-shirts, and discovering his name and picture emblazoned on the side of the BAC 1-11 was something he would never forget.

Andy captured the moment of Murray receiving his guard of honour, but after take-off, the fun really started. As soon as the seatbelt signs were turned off the cabin erupted into one big party, with champagne being served by one Murray Walker OBE, dressed in the European Aviation cabin crew uniform of blouse, scarf and skirt! Eventually Andy gave up trying to film the proceedings, knowing that none of it could be used. The whole flight was by far and away the most memorable F1 party I've experienced in nearly 30 years, and it was topped off by Stoddart, on landing back in the UK, opening the forward door when it was still set to 'automatic'. That popped the emergency slide, and he then beckoned all of us – including Murray, hip replacements notwithstanding – to disembark the aircraft by jumping on to the slide and bouncing down to ground level. I like to think Paul gave the airline mechanics who had to re-package and stow the emergency slide that night the next day off!

Then, all too soon, came the United States GP at Indianapolis, Murray's last race as commentator. This had been his choice, as he felt it might have been somewhat of an anti-climax had his last race been the final round of the season in Japan, which traditionally, due to the time difference, doesn't attract as large an audience as the US races. North American races are generally shown at prime time in the UK, and as everyone knew it was to be Murray's last race, early evening would make for an easier time for people to watch.

It was an emotional weekend. Our production manager, Sally Blower, helped to organize a surprise farewell event for Murray on the Saturday evening after qualifying. The event was being held in the Paddock Club, Formula 1's VIP hospitality area, but no one wanted Murray to twig what was going on, so we all sneaked up the back stairs to the club. When Murray was brought in through the main entrance, he was stunned to see so many people from the F1 paddock: all the ITV crew, other members of the media and key people from the teams, including most of the drivers. Bernie Ecclestone was there, along with Michael Schumacher, David Coulthard, Mika Häkkinen, Jenson Button, Juan Pablo Montoya and many other famous names. Host Tony Jardine did a charming review of some of Murray's most famous quotes and 'Murray-isms'. He asked the drivers to come up and read theirs out with Murray standing alongside. Indianapolis Motor Speedway owner Tony George presented Murray with one of the bricks from the famous original track. It was all done with a huge amount of affection.

Race day came, but as the Grand Prix was the first big sporting event held in the USA after September 11 the tone was subdued.

It wasn't a particularly significant race. While Mika Häkkinen won for McLaren, Michael Schumacher had wrapped up both the drivers' and constructors' championships three races earlier in Hungary. Almost before I realized it, we had come to the end of the show. I remember Murray's last sign-off as vividly as if it were yesterday. There were four of us in the commentary box; Murray, Martin, cameraman Mat Bryant and myself. With the post-race analysis all done, at around 9.30pm UK time, Jim Rosenthal handed back for Murray to say farewell, and to take the programme off-air. And in typical Murray fashion, he kept it straightforward and direct. 'That's the last from me,' he concluded. 'All I can say is that it always has been a pleasure, and I hope you will enjoy Grand Prix racing from now on. Goodbye.' That was it.

The director cut to a package we'd prepared, a musical tribute edited to the Cole Porter song 'You're the Top'. Murray put down his microphone and had a little hug with Martin. He then quietly gathered up his stuff, stowing it in the vintage 'Shell Oils Grand Prix of Europe' bag that he carried everywhere, left the commentary box, and walked down the stairs with Martin. There weren't any tears, it was just a very matter-of-fact sign-off. It was obviously a moment of huge significance for him, and for all of us, but unlike Murray we would be going on to the season finale in Japan. Murray got a car back to the hotel, stayed Sunday night, just as he had done that evening in Brussels the previous year, and flew home the next day to get on with the rest of his life.

Over the next couple of years, I wondered if Murray had retired too soon, because he showed no signs of being done with F1, or of being ready to stop working. He signed up to be a Honda ambassador in 2005, providing their guests with paddock insights and first-hand stories about his time in F1. He went to quite a few races in a Honda shirt and cap, which always felt a bit strange, but it was lovely to see him around and enjoying life. He was happy to remain involved in F1. He told me he feared that had he stayed at home and sat and read books, or gone cruising around the world with Elizabeth, he would have lost his mental sharpness. And as it turned out, the commentary box hadn't seen the last of Murray because he came back for one final F1 commentary at the 2007 European Grand Prix at the Nürburgring, when he filled in at BBC Radio 5 Live for David Croft, who was attending the birth of his son. That really was the last one for Murray, performed at the grand old age of 84. He had an extraordinary life and career in the sport he loved so much. I'll be forever grateful for the time I spent with him and to this day there isn't a race weekend that goes by that I don't think of him, and hear the echo of that voice of his, reverberating around the race track.

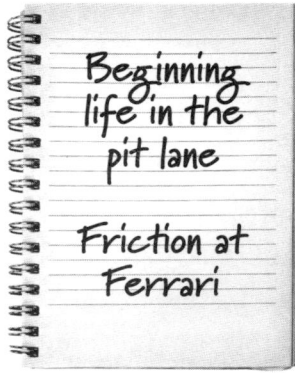

Chapter 6

Learning the Ropes

It's hard to think of any other sport where one person was for so long its voice and effective figurehead. Murray Walker's professional longevity meant a 50-year wait for the job of commentator to come up. However, at the end of 2001 that opportunity finally arose. Waiting in the wings was James Allen. Deputizing for Murray alongside Martin Brundle three times during the 2001 season, James had proved himself a perfect fit for the role and for the viewers was a known and trusted voice. Thus, he was duly promoted from pit lane reporter to the role of ITV F1's lead commentator. All sorted, then. Apart from one detail – who was going to replace James as pit lane reporter?

It is, as you're hopefully learning from this book, a job that requires a very particular set of skills. James had successfully defined the role and I had watched and learned as he did it. I was already part of the team, had reporting experience from my radio days and, by now, a decent spread of F1 knowledge, so I made the shortlist to replace him. The other frontrunner, Kevin Piper, had also been part of our production since the start. He had previously been Head of Sport at Anglia TV, and he was a highly experienced and accomplished journalist. Grand Prix racing had a long and distinguished history in the Anglia region. Lotus was based in Hethel in Norfolk, not far from the Snetterton circuit, where local boy Martin Brundle had sparred with Ayrton Senna as they both learned race craft competing in F3, and Kevin, or 'Pipes', as he was widely known, had reported regularly on F1 stories over the years. Before joining Chrysalis Sport to produce the F1 shows for ITV, Kevin also had the honour of playing himself in the opening scene of the football comedy film *Mike Bassett: England Manager*.

In 2001, while James Allen was trialling to take over from Murray, Kevin and I each had an opportunity to try out for the pit lane reporter's role. My first chance came at the Brazilian Grand Prix. It was more than a little nerve-wracking. Starting as I was expected to go on, my first ever report was live and broadcast instantly to millions of people in the UK and around the world. The subject was the ever-changeable São Paulo weather and, in particular, as I gamely pointed to it on camera, 'that cloud, over there'. In the end, the cloud in question wasn't rain-bearing but the threat of it had been enough to affect the teams' strategies. What I was attempting to do was convey the mood of tension among the teams, and communicate the possible consequences of the rain that never fell.

This and subsequent contributions from me must have not been entirely awful because at the end of the year Neil Duncanson recommended to ITV's Brian Barwick that I be given the pit reporter's role full-time. And that was it. I've been in the job ever since!

With the exception of some early risers who might have caught me on Capital FM's breakfast news bulletins, I was completely unknown to a British audience. Barwick gave me some advice: to play myself in quietly, not to try to make a big splash, to do the basic things consistently and well, and to give myself time to become established in the role. Helpful, but by far the most useful advice I received was from the previous pit lane reporter, James Allen himself. 'The best reports from the pit lane essentially drop little golden nuggets of information into the race commentary, and then allow the commentators to get on with it,' he said. 'Viewers can't always hear every word you say from a noisy pit lane, so keep it simple, be precise, get the information out there, and then shut up and focus on anything else going on in the pits that the commentators can't already see for themselves.'

Murray Walker also threw me some advice: 'Inform and entertain' was his golden rule. His feeling about the art of live reporting was that it's as important to think about the way you say things as the detailed content. He was right. You can take your time finessing pre-prepared packages until they are frame-perfect, but a reporter will always be judged on the quality of their contribution to the race and qualifying commentary. So the trick was firstly to make sure it was informative, and secondly, if possible, entertaining. But of all the advice I received, perhaps the most memorable came from Neil Duncanson, who simply told me to 'be a sponge'. I knew what he meant. Even though I'd been coming to the races for five years, and knew my way around, I also knew that to really delve into the details I'd need to fully immerse myself into the world of the pits, paddock and media centre.

'Stay out of the TV compound,' Duncanson warned. 'You'll never learn anything from broadcast trucks and the production office. Get out into the paddock. Make friends and contacts. Talk to drivers and engineers, team principals and FIA officials. Get a feel for their backstories and motivations, find out what they're thinking, and how they're approaching the weekend.'

Neil's advice is as valuable now as it was back then. Indeed, it's still how I approach every race weekend. But while I was thrilled with my new role, in terms of what was happening on track, my first season as pit lane reporter was proving to be less than thrilling – Michael Schumacher's Ferrari was dominant and he was winning practically every race. Until, that was, a sunny May day in Austria, when I had my first big F1 controversy to report on.

2002 was no flash in the pan for Ferrari. With hindsight it's clear they were in the middle of the five-season-long stretch of dominance that ran from 2000 to 2004. Michael Schumacher was the team's designated number-one driver. Early in the year there had been flashes of pace from a resurgent Williams team and their promising BMW engine, but Michael had won four of the first five races, and round six in Austria found him 21 points clear of his nearest rival. The feeling in the paddock was that Williams weren't quite ready to win consistently, and that Schumacher was a solid bet for the title. Ferrari had just re-signed Michael's teammate Rubens Barrichello, providing stability on their side, while Barrichello was pleased to have the security of a two-year deal.

Rubens didn't so much wear his heart on his sleeve as carry it around like a giant helium balloon. You always knew immediately what he was thinking, largely because he'd tell you. From tears of joy at winning to tears of despair when things went wrong, over an 18-year, 326-race career, Rubens did a fair bit of crying. But he wasn't just emotional, he was also honourable.

His reasoning in renewing his deal with Ferrari was simple. Would it be better, he asked himself, to be a number-one driver in a less competitive team, driving a car that wouldn't necessarily allow him to win races, or to be a 'team' driver at Ferrari where he did in theory have a chance to challenge for the world championship? Of course, to do that he'd have to beat Schumacher. That was a big ask, but Rubens had proved himself an excellent driver and knew that it wasn't impossible. From Ferrari's point of view, the number-one and number-two philosophy began with a simple concept – both drivers started the season on zero points. Whoever scored the most points early on and therefore showed themselves as most likely to win the world championship would become the team's focus, and the other driver would take a supporting role. In Schumacher's previous seasons alongside Barrichello he had taken the lead-driver role by the time the European season started, or at least by the time of the Monaco Grand Prix at the end of May, and then as a result had enjoyed number-one status for the rest of the season with Rubens acting as his rear-gunner, racing in the team's – and Michael's – interest, which included obeying team orders when they were issued.

The more cynical in the paddock believed Ferrari would find ways to make sure that Schumacher was the points leader by the time the European leg of the championship started, whether that was down to a car advantage over his teammate, favourable race pit-stop strategies, or some other preferential treatment.

People even went so far as to suspect that Michael had these advantages formally included in his contract with Ferrari, something that he always firmly denied.

All of this notwithstanding, Schumacher certainly had the better start to the 2002 season, winning in Melbourne, São Paulo, Imola and Barcelona. By complete contrast, Barrichello's campaign had started disastrously. He retired from the first three races and failed even to start the Spanish GP from his second place on the grid, thanks to a gearbox failure on the formation lap. In his darkest moments of paranoia Rubens might indeed have imagined that these misfortunes had been created by Ferrari to ensure that Michael could rack up the points untroubled, satisfying the supposed conditions of the number-one driver. If Barrichello suspected that, he never said it publicly, and even the lightest investigation of his technical issues confirmed that they were simply a run of bad fortune. This was recognized by the Ferrari team principal Jean Todt, who sympathized with his travails, and appreciated his loyalty in his reluctance to blame the team. Todt had rewarded Barrichello with the early contract extension, and the Brazilian flew into Austria full of confidence and ready to reboot his championship challenge.

The Spielberg track, a shortened version of the original Österreichring built in the late 1960s, was and remains a stunningly beautiful and technically challenging venue. Some drivers always went well there, some did not. Schumacher was in the latter camp, having never previously won in Austria. It was one of the very few tracks where his ability to throw the car into corners and then sort out the sliding rear end was rendered fruitless, as the circuit layout rewarded a more precise approach and favoured Barrichello's driving style.

The Brazilian led away from pole position, and for most of the race he ran comfortably ahead of Schumacher and a chasing Juan Pablo Montoya. But rather than just sit back and let the race play out, at around half distance Jean Todt got the frights. Worrying was something that Todt did quite a lot of during races. I often saw his fingernails wrapped in medical tape to prevent him biting them to the quick. Even though Schumacher's lead in the drivers' championship was a healthy one, experience told Todt that things could change very easily over the rest of the season. All it would take was some of Barrichello's mechanical unreliability to strike Schumacher's car and the German's lead could be wiped out. Whether those concerns were well founded or not, Todt was in charge, and he wasn't about to let Barrichello take four points off Schumacher, no matter how much Rubens deserved the win.

He duly instructed Ferrari's technical director Ross Brawn to execute the driver swap so that Schumacher could take the full 10 points (the score for a win at that time) and further extend his lead in the world championship standings. Ross in turn told Barrichello's engineer Gabriele Delli Colli to order his driver to give up the lead. Team orders were nothing new and were an established part of motor racing at this time – it was just another routine race decision at Scuderia Ferrari. But someone wasn't listening. For lap after lap, Barrichello ignored the instructions from the pit wall, and stayed in the lead. Brawn even tried talking to him directly – with no response.

Spielberg is a short circuit, so to complete the mandated 305-kilometre Grand Prix distance, the drivers had to drive a total of 71 laps. Again and again, Rubens passed by the pits, retaining his lead. I was watching the drama on the Ferrari pit wall play out in real time. This was before the current era of fully open team radio channels, so as broadcasters we had little idea of what was going on from the drivers' point of view, but what I could see from the pit lane was a highly unusual amount of back-and-forth communication between Todt and Brawn. Notes were being pointedly passed across the carbon fibre-lined desk, and there were intense discussions between Brawn, Delli Colli and Todt. Conscious of eyes and TV cameras upon them, they were just about staying outwardly calm, but one person wasn't playing ball.

Ten years later, in an interview with Brazil's TV Globo, Barrichello described the closing stages of the race. 'It was eight laps of war. It's very rare that I lose my temper but I was screaming on the radio. I kept going to the end, saying I would not let him pass.' Whether Barrichello's reasoning for disobeying the team order was that he felt the swap was unnecessary given Schumacher's championship lead, or that he thought it was simply unsporting, didn't matter. On the radio, Barrichello said he was told that he must 'comply or face stiff consequences.' Looking back on it, Rubens referred to a 'broad form of threat', but this was not revealed by the F1 TV camera now trained on Todt. What we did observe was Todt's final, slightly desperate radio call which was played out on the main TV feed. 'Rubens – last lap. Let Michael pass for the championship. Let Michael pass for the championship, Rubens, please.'

At the last minute, Barrichello complied. To make it absolutely clear this was a team order, and at what was clearly the worst possible moment for Ferrari in PR terms, Rubens lifted his accelerator coming out of the last corner, allowing Schumacher, close behind him, to power past to the finish line. Unlike Lando Norris, who years later would do the same to let Oscar Piastri win the Qatar GP sprint race in 2024, this wasn't Barrichello's choice – and loud boos rang out from the crowd, objecting to the apparent manipulation of the race result.

What really made Ferrari's decision baffling for most observers was that while it was in line with the way the team generally raced, they hadn't needed to do it. Any reasonable projection could see that Schumacher was likely to win the world championship easily, and in no way needed the extra four points that separated first from second. Allowing Barrichello to take his first win of the year would have given him a huge confidence boost after his poor start, and that would in turn have helped the team overall in its battle for the constructors' title. Instead, Ferrari was now in the firing line, with even the most loyal fans upset by what they had witnessed.

With the negative reaction from the Austrian crowd ringing in his ears, Michael was apologetic in the post-race interviews. Barrichello by contrast had already made his point in the most public way, so was actually pretty relaxed about the result. In the moments before the podium ceremony, it was down to me to make some sense of Ferrari's decision for ITV's viewers. Ross Brawn stepped off the pit wall into a throng of reporters. I had positioned myself second in line behind home broadcaster ORF. Ross was besieged and looked slightly taken aback as to why the result had caused such controversy.

He turned to me. 'Rubens had won the race, hadn't he?' I demanded. 'Why did you swap?' Ross conceded Barrichello had the race effectively won. 'But in the interests of Ferrari and the drivers' championship,' he said, 'we made the decision.' Then came the nugget of information unavailable to us throughout the Grand Prix, an added detail from Brawn that changed the story. He revealed that Michael hadn't been racing Rubens at all, and that in fact, such was their dominance over Montoya's Williams the drivers had been instructed to back off, turn down their engines and manage their pace, coasting to the end – both cars were merely on a cruise to the finish. This raised two additional questions that we would go on to debate long into the evening. If the Ferrari drivers were so much quicker than Montoya, then why was Todt so worried about Williams closing up in the championship? And, second question, if both had been allowed to push flat-out, would Schumacher really have been able to outpace and overtake Barrichello – something he had been unable to do all weekend?

Brawn was being scrupulous in his honesty, perhaps reasoning that if people understood the cars hadn't really been racing they might mind less about the team orders – but his statement mixed uncomfortably with the result, and viewers were left wondering, if there hadn't actually been a race, what exactly it was that they had been watching for the last hour and a half? Furthermore, by switching positions at the most awkward and embarrassing moment, only yards from the finish line, Barrichello had most publicly signalled his frustration with the decision, which suggested that all was not rosy in the Ferrari garden.

Things were about to get much less rosy still on the podium. Embarrassed by the crowd's hostile reaction, and keen to offer an olive branch, Michael refused to take his place on the top step, insisting instead that Rubens take the winner's position. The move was taken in good faith by the local dignitaries handing out the trophies, but the FIA was less impressed and would go on to summon both the team's representatives and drivers to the World Motor Sport Council at their Paris headquarters. Under the gaze of the camera crews lined up outside, in walked Todt, Brawn, Schumacher and Barrichello to face the music, but the problem for the FIA was that Ferrari hadn't actually broken any rules in switching the positions – team orders have a long history in Grand Prix racing, and at that time they were not outlawed.

What the World Motor Sport Council was able to penalize, however, was the breach in protocol when the drivers had switched positions on the podium. In the FIA's eyes this had 'confused and embarrassed' the podium dignitaries (a bit of a stretch given they are recorded on TV smiling and applauding Schumacher's gesture), but it was nevertheless against the rules. Ferrari avoided a points penalty in place of a $1 million fine, half of which was suspended against future good behaviour. The FIA also decided to make a stand in order to deter future team-order controversies, and from 2003 the practice of F1 teams instructing one driver to let the other pass for a race position was banned.

The rule lasted until 2011, when it was quietly dropped, in essence because the teams had found so many ways of subverting it, most obviously if it was the driver's decision. The most famous example came at the 2010 German GP, once again involving Ferrari. Felipe Massa was leading teammate Fernando Alonso when his engineer Rob Smedley sent a highly unsubtle coded request for a swap of positions. 'OK,' Smedley began with audible reluctance, signalling that it was a decision he was being made to execute. Never has an 'OK' done so much heavy lifting. After a beat, Smedley continued. 'Fernando is faster than you. Confirm you understood the message.'

Realizing that he had just been designated the support act at Ferrari, Massa duly let Alonso through. Smedley's subsequent message to Massa of 'OK mate, good lad. Just stick with him now, sorry', provided a further indication that the order had come from the team. Alonso went on to win the race while Massa fended off a charge from Red Bull's Sebastian Vettel to finish second.

This set of radio messages once again landed Ferrari before the FIA race stewards, accused of breaking sporting regulations, and the team was again found guilty and fined. A later investigation by the World Motor Sport Council upheld the stewards' decision and the fine, but stopped short of taking any further sporting sanctions. Behind the scenes the WMSC directed F1's Sporting Working Group (a committee made up of F1 team managers and FIA officials) to take another look at the regulation banning team orders due to it being unenforceable. And so, in 2011, eight years after it had first been brought in, article 39.1 of the FIA Formula 1 Sporting Regulations banning 'team orders which interfere with a race result' was deleted from the statutes.

As for me, the aftermath of Austria 2002 gave me reason to consider the dangers of bringing one's feelings about a racing decision into an interview. Brawn could tell by the tone of my questions that, like most people watching, I felt that his and Todt's decision had been unfair, unnecessary and unsporting. At the end of our interview in the Spielberg pit lane I had turned away from the mêlée surrounding Brawn to allow another reporter in to ask their question. It was only at the following race in Monaco when I approached Ross for a post-qualifying interview that I discovered he was annoyed.

'Oh, so you want to talk to me now, do you?' Brawn said, with a trademark raised eyebrow.

'Yes, I always want to talk to you, what do you mean?'

'Well, you didn't seem too keen on listening to my answer or concluding the interview when you just turned and walked off on me in Austria.'

It belatedly dawned on me how it must have looked to Brawn at the time. As I turned away the TV audience would have heard me say into my microphone, 'Well thanks for explaining, Ross' or something along those lines, but crucially Ross hadn't heard it. Viewers also couldn't see Brawn's point of view as he watched me duck away from him and out of the scrum. That taught me the important lesson that you should always end any interview politely and directly, thanking the interviewee to their face. And if you can avoid giving them any indication as to whether you agree with them or not, then so much the better.

Michael Schumacher won the 2002 world championship with a massive total of 144 points, almost double Barrichello's 77 and nearly triple Montoya's final tally of 50 points. Jean Todt needn't have got the frights after all.

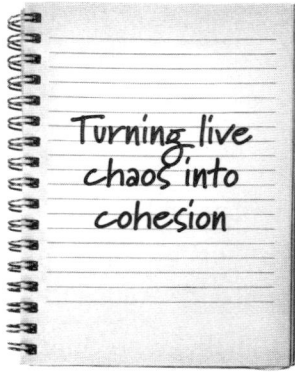

Turning live
chaos into
cohesion

Chapter 7

Dealing with Disaster:
The Pitfalls of Live Broadcasting

———

Hang around behind the Formula 1 garages for long enough with a TV camera and you're guaranteed to hear one question: 'Are you live?' Most frequently asked over a Grand Prix weekend by reporters to other reporters, and by mechanics, engineers, team press officers and drivers, it essentially means, 'What exactly is going on here, with this TV camera?' They want to know whether the moment is being beamed out to millions, or whether they can afford to be slightly more relaxed, the equivalent of standing to attention or standing easy. In my case, the answer is almost always, 'Yes, we're live.'

It's not just me. F1 has around 35 broadcasters at the circuit presenting all or some part of their programmes live. There are others who are there with film crews, but who are not live, a good example being the crew filming the Netflix documentary series *Drive to Survive*. Netflix aim to try and capture as much as they can, editing it all down to the most compelling stories once they've figured out the story arc for the season, but what that means is that most of what they film will never be seen or heard.

For F1 personnel it's important to know who's live and who isn't. If there is any professional code among the F1 paddock media, it's that the broadcasters who are going out live, with no opportunity to stop or edit, are afforded a higher priority than those who are not. Of course, the non-live (recorded or 'off-tape') camera crews still need their interviews, so an interview scrum, whether around a driver or team principal, usually comes down to a ballet of split-second timing to make sure that everyone gets what they need, and nobody is left with a hole in their programme. One final point to note is that it's generally good policy to let the drivers know before an interview starts whether they're live or not, so they don't swear (although if you make a point of mentioning it to Australia's Daniel Ricciardo he won't miss a beat before replying with a swear word just for everyone's amusement).

Getting any live broadcast to work takes a huge amount of preparation and, given that triumph or disaster is only ever a split-second away, broadcasting F1 is exceptionally demanding. It's the typical swan paddling on a lake scenario. Things might look serene from the surface, but out of view under the water everyone is working furiously to make the programme happen. To me, live TV is a bit of an art form. There is so much that could go wrong at any point, it's remarkable we don't have more disasters than we do.

When plans go out of the window, it's usually because of circumstances beyond anyone's control. The most familiar culprit is the weather. Every once in a while rain means we get a delayed or interrupted session or race which leaves us trying to fill the air time, grateful for shots of marshals dancing in the rain, or – as

I remember at Suzuka, Japan, in 2017 – mechanics floating paper boats they had made down the flooded pit lane. Under more extreme circumstances the qualifying session might be abandoned, postponed until Sunday morning, or even a race time suddenly moved forward to avoid an incoming storm, as happened at Interlagos in 2024. As broadcasters we know a rainy weekend usually leads to good TV, as racing in the wet tends to bring out some amazing performances from the very best drivers. Memorable wins for Lewis Hamilton at Silverstone in 2008 and Max Verstappen in Brazil in 2024 are just two that come to mind, races where in each case in the wet, these drivers were simply in a class of their own.

But inclement weather can also lead to weekends where things go very wrong indeed. One such was the Japanese GP of 2014. We'd been warned of a typhoon approaching the Nagoya Bay area, near to where the Suzuka circuit is situated. Typhoons aren't uncommon in Japan, especially in the autumn, and so, despite forecasts that the weather was due to get worse over the course of Sunday afternoon, there had been no adjustment to the schedule, and no amendment to the race's 3pm start. F1 boss Bernie Ecclestone, who wasn't actually present at the track that weekend but was in touch by phone, was reluctant to interfere with the broadcast slots scheduled around the world, and the Japanese organizers had made the decision to go ahead, believing that they could get the race completed before the worst of the rain reached land.

Conditions were tricky and wet as the race went ahead. As the daylight began to fade, it looked as if the 53-lap race would be completed without serious incident. But then, on lap 41, Sauber driver Adrian Sutil spun off into the barrier on the outside of the

Dunlop Curve. It wasn't a particularly hard impact, and it was quickly clear that Sutil was out of the car and OK. Race control told the marshals to wave double yellow caution flags as the recovery workers dispatched a small crane to move the stranded car. Double waved yellow flags were standard practice and generally considered sufficient for this type of incident. The rules stated that drivers under waved yellow flags had to reduce their speed and be prepared to stop.

Then, at the very same spot where Sutil had spun off, Jules Bianchi lost control of his Marussia. He slid across the wet grass at high speed and hit the heavy counterweight of the recovery crane with such force that it jolted up momentarily. The incident wasn't shown on the world feed, and initially we were confused as to what had transpired. The world-feed director would have seen what happened, but the general principle is that serious accidents are not shown until they are sure the driver is alive and well. When the director did cut to a distant shot of the scene, we could just about tell through the murk that a second car had gone off at the position where the crane was working on Sutil's car. The Marussia wasn't clearly visible, but the activity of the marshals and medical team did not bode well.

Furthermore there were no replays – a sure sign that something serious had happened. Realizing this, David Croft immediately shifted the tone of the commentary down several gears. Our worst fears were confirmed when, after the cars had circulated for a couple of laps under the safety car, there was a red flag, and the race was stopped.

Our first job as broadcasters was to try to figure out what had happened so that we could calibrate how we were going to deal with covering it. Everyone watching the live coverage was looking for news about Bianchi's condition. His team told us that they hadn't received any response on the radio, so at that point they didn't know anything more than we did.

Jules Bianchi was a young Frenchman who had seemed to have it all – incredible skill in a racing car, a great personality, movie-star appearance, and a bright future ahead of him. His parents were friends with the Leclerc family, so much so that Bianchi had been named godfather to the young Charles Leclerc, who in 2014 was racing in Formula Renault. Bianchi was generally regarded as one of the future stars of F1. A protégé of Ferrari, he was in his second year with Marussia, and was set to move up to the Sauber team in 2015. Ferrari planned to nurture him in such outfits that ran Ferrari engines until a place opened up for him at the main team.

There had been many moments where Bianchi had demonstrated that promise, none more so than his eighth-place finish at the 2014 Monaco GP. The aftermath of that race particularly stands out in my memory. The pit reporter's job is to quickly interview the winner and then look for what's going to be the second story of the post-race programme's recap, most regularly a surprise placing or a feel-good result from someone in mid-field. In this case Bianchi had scored the first points for both himself and the Marussia team. I headed down to the garage, where I found the mechanics hugging each other in celebration. Team patriarch John Booth was walking around, slapping everyone on the back. It was a joyous celebration, because it meant the team would move off the back of the constructors' championship and begin to earn prize money that would guarantee their survival in F1.

The contrast with the tense atmosphere post-race at Suzuka just a few months later could not have been more marked. There were a lot of rumours going around, and everyone was hungry for accurate information. It was Matteo Bonciani, then the FIA's head of communications, who eventually came out to speak to us. He was met with an impatient rush of microphones and cameras, and I felt I had to implore my colleagues to step back and give Bonciani some space. It was a very intense moment. Bonciani gave us a prepared statement to the effect that Bianchi had sustained a head injury, and that he had been transferred to hospital in Yokkaichi. And there was simply no more news to give us than that.

In the case of information from the teams or the FIA, it's important to focus on facts and not speculate. There's not much that can stop the internet, however. In this case, videos taken by spectators at the Dunlop corner began to emerge on social media that gave us the best indication yet of just how the accident had happened and how serious it really was. We had the statement from the FIA, we had limited information from the team, and from the pictures online we had a pretty good idea of the nature of the accident and the injury Bianchi had suffered. The crash was a stark reminder of how exposed the drivers really were. We still had limited official news at the time we went off-air, which is where our live element ended and the pre-recorded programmes picked up. It was a very sombre flight back to London.

After several months in a Japanese hospital Bianchi was eventually transferred back to France. Sadly he never regained consciousness from his brain injury, and passed away in July 2015, aged just 25.

In the immediate aftermath of the accident the FIA announced an investigation. Under scrutiny was the speed at which Bianchi had been driving through the corner under the double yellow flags when he lost control, but there was equal scrutiny of whether more could have been done to prevent the circumstances leading to the incident, and quite rightly the sport learned lessons, as is so often the case following a serious accident. The investigation led directly to the introduction of the virtual safety car – much like an average speed zone on a motorway, a read-out on the dashboard forces all drivers to reduce their speed equally, and removes the discretionary element following an incident. It was trialled over the remaining races of 2014, before becoming a standard procedure, particularly when marshals are engaged in recovering cars from the track.

Bianchi's crash also added impetus to the FIA's research into a head-protection system for F1 cars. Various options had been under development, including a laminated polycarbonate windscreen – something that was later adopted by IndyCar – and two metal bars either side of the driver's head, aimed at deflecting large objects like wheels. There was a fair amount of resistance to the concept. Some drivers spoke of their concern that F1 would lose the essence of being an open-cockpit category, while others didn't like the idea of having their vision partially obscured.

The FIA would eventually settle on what became known as the Halo, a cast titanium wishbone-shaped device that is mounted either side of the driver's cockpit and secures to the front bulkhead of the chassis just in front of the steering wheel. Opinions were mixed when it was introduced at the beginning of 2018. But just like the HANS device – the Head And Neck restraint System that was made mandatory in F1 a few years earlier – the drivers quickly got used to it and recognized its value. Since its introduction the Halo has certainly saved the lives of drivers, not just in F1, but in the other formulas in which it was adopted. Perhaps the most notable example was another accident where we were left shocked and desperate for information – but which ultimately had a much happier outcome.

Having started the 2020 Bahrain Grand Prix from near the back of the field, Romain Grosjean was clipped by Daniil Kvyat's Alpha Tauri on the straight that followed Turn 3. His Haas speared right, slammed into the Armco barrier at high speed, and exploded in a ball of flame. The chassis punched right through the metal barrier, snapping the rear of the car housing the engine and gearbox clean off. It was a shocking, sickening crash.

As with the Bianchi accident, the F1 world-feed director quickly cut away, and there followed an agonizing period during which neither those in the pit lane nor viewers around the world knew what had happened. After around 30 seconds, we finally saw shots of Grosjean. Not only had he made it out of the car but he was being walked to the medical car. Replays showed how he had twisted his body and pulled himself up through the Halo device as the fire blazed all around. He was then helped over the barrier by the FIA medical delegate Dr Ian Roberts while medical-car driver Alan van der Merwe sprayed them both with fire extinguishant. Grosjean suffered burns to his hands but the fire-resistant race suit, balaclava and underwear did their job. It was a miraculous escape, but the Halo device had been crucial. Without it Grosjean would surely have suffered a severe head injury on impact with the metal barrier and have been helpless to escape the subsequent fire.

Later that evening, after the race had finished and we knew Grosjean was safe and well in the medical centre, my producer asked me to go down to the crash site to file a report on what the scene was. The exit of Turn 3 was easily accessible from the paddock, and just within the range of our microphones and camera.

Grosjean's wrecked chassis had long since been transported away, but the first thing that struck me most forcefully was the smell of burning plastic and carbon fibre hanging in the air. It was a sobering sight to see the discarded sections of damaged barrier, temporarily replaced for the restart by a section of concrete wall. Seeing the aftermath in person made it even more incredible to think that Grosjean had been able to survive such a horrendous accident.

Such are the perils of live broadcasting. Motorsport is inherently dangerous, and the speeds involved are so high that any accident has the potential to be fatal. It gets safer every year as lessons are learned from incidents and near misses, but the nature of F1 is its unpredictability. Just as the drivers accept that risk when they put their helmet on and step into a car that in the wrong circumstance could kill them, so we also accept the risk when we turn on our cameras and broadcast live TV.

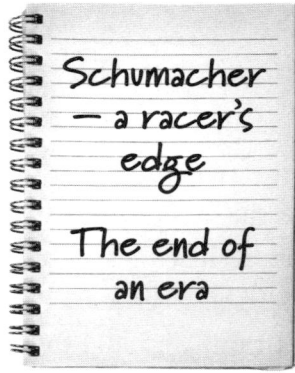

Schumacher
— a racer's
edge

The end of
an era

Chapter 8

Michael

It's a pleasant May evening in Monaco and journalists, reporters and camera crews have gathered outside the Ferrari motorhome a few hours after qualifying for the 2006 Monaco Grand Prix. Earlier that afternoon we had witnessed Michael Schumacher shunting his car into the barrier at the penultimate corner, La Rascasse, causing yellow flags to be brought out and wrecking his championship rival Fernando Alonso's final chance of clinching pole position. This is motor racing, and accidents can and do happen. This one, however, had a distinctly dubious look to it – the way Michael's car ended up just short of damaging itself against the barrier yet still effectively blocking the track had suggested to many observers that this wasn't an accident at all.

Schumacher was sitting in provisional pole position. Knowing Alonso had the chance for another hot lap, on his final run Schumacher, a driver renowned for his finesse and skill, seemed to clumsily wrestle with his car at the Rascasse corner until it came to a stop, blocking half the track and forcing anyone completing a lap behind him to slow down and abort. Alonso was one of the drivers affected, and as the session ended Michael had held on to his pole. I had been listening in to the commentary from my regular Monaco position – wedged between a stack of tyres and a marshals' cabin – and heard the drama unfold. I then walked down the pit lane accompanied by cameraman Andy Parr.

The first person we encountered was Alonso's team principal at Renault, Flavio Briatore. Schumacher was close to Briatore – they had won two titles together in the Benetton days – but on this occasion the team boss was furious with his former driver. The charismatic Italian practically had steam blowing out of his ears. He declared himself in no doubt that Schumacher had faked the accident to guarantee himself pole and to deny Alonso or anyone else the opportunity of beating him. 'This is the way Ferrari manage, you know,' he fumed, before ending our interview by stalking off to prowl the garage. He wasn't alone. The FIA race stewards agreed that the incident looked suspicious. They called Michael and his team representatives to a hearing, and examined closely the telemetry data from Schumacher's car.

As their deliberations dragged on all anyone could do was wait, and outside the Ferrari motorhome seemed as good a place as any. Ironically, waiting for Michael Schumacher was not something I was used to over the six previous seasons covering his Ferrari career. One of Michael's many qualities was his immaculate timekeeping, a consequence of his famous self-discipline which extended just as much to media time as it did to physical or technical preparation. He was always very clear that his two top priorities were his fitness training and driving. Of course, he had other commitments, so to make sure he could focus on his priorities he ensured that anything else fitted exactly into its allotted time. He was helped in no small measure in this by Sabine Kehm, a Berlin-based journalist who joined Michael's management team in 1999 as media consultant and still works for the Schumacher family today. From the beginning of the race weekend Michael liked to stick to a routine. He'd always arrive at the circuit at the same time, just before lunch on Thursday morning. Meetings with engineers would follow before his media commitments, always at 4pm. He would allow time for some fitness training at the end of the day before the track action began on Friday morning. Time was his most valuable asset, and Michael guarded it carefully.

When it came to TV interviews, an activity he was obliged to do but which didn't make him or his car any faster, Schumacher was direct and to the point. James Allen had mastered the art of the Schumacher interview and gave me some tips. As he explained, most drivers get bored with always answering the same questions, such as 'What are your targets for the weekend?' or 'How do you expect the car to suit this track?' Most will just (figuratively) roll their eyes, give routine stock answers and move on, reasoning this will get the whole thing over and done with as quickly as possible. Not Schumacher. If he was obliged to answer questions, even in a brief TV interview, he demanded just as high standards from the interviewer as from everyone else in his professional life. And if you didn't demonstrate that you'd put in the effort to rise to those standards, he would give you a short and basic answer. You'd be left with practically no material for your news report and be in little doubt that he didn't think much of you or your question.

Allen advised me to take extra time to come up with questions for Michael that were intelligent, considered and original – specific angles that would make him think. This way he would perhaps get something from the interview too. I took that counsel very seriously, and for my first few encounters with Schumacher in 2001, I made sure that I asked the kind of question that made him ponder his answer. He had a bit of a tell when you'd asked a decent question – he'd purse his lips, look up to one side and then give a full response rather than just put the question away with the dreaded couple of words. The trick with the pithy questions was to leave them till the end of the interview. Michael, knowing the score, would usually oblige with a quick soundbite and then a wink at the camera. Generally he responded well to my lines of questioning. And before too long, I noticed camera crews from other broadcasters would come and stand next to me when I was about to interview Schumacher in the group media pen. One weekend I asked Fuji TV cameraman Ollie Parnham what was going on.

'Well, he seems to like you,' he explained. 'And you get good answers. So if Kaz [Kawai, legendary Japanese F1 broadcaster and my pit lane counterpart] isn't on site, we know we'll get what we need if we just pick up your interview.'

I was relieved that Michael hadn't decided I was a complete idiot at the start of our professional relationship, and I certainly respected him. We did get to know each other over the years despite the fact that we were never formally introduced. I assumed he wasn't particularly interested in who was asking the questions, so I hadn't wanted to waste his time by introducing myself when I first got the pit lane job. I had the feeling a quick 'hello' would suffice when he appeared for our first interview. For those of you taking notes on how to do the job of pit lane reporter, I'd mention here that there's always a side benefit of such a greeting at the beginning of any F1 interview, as it gives the camera operator time to press the record button and check that the focus is sharp before we begin. It also allows the interviewer to establish a kind of structure and take charge of the situation, something that Schumacher was keen on. Understandably he didn't like the chaos of a scrum of reporters and cameras, not least because he'd been in the middle of quite a few of them over the years. And at around 10pm on that Saturday in Monaco, he was about to be right in the centre of another.

The stewards were taking their time to come to a decision on whether Michael's accident was deliberate. Typically, in FIA stewarding, if it is decided that a rule has not been breached, or if it has and the decision is straightforward, any penalty will be handed down reasonably quickly. However, cases where the stewards must consider the kind of intent that will reflect badly on a competitor are naturally highly sensitive, resulting in longer deliberations to make sure that the outcome is the correct one. In Monaco, those deliberations entered their fifth hour while Schumacher waited for news in the Ferrari motorhome. Meanwhile, across Monte Carlo every other F1 driver was trying to dodge the Saturday night parties in search of a quiet dinner and an early night.

In his book *Edge of Greatness* James Allen tells the story about how Mark Webber and Fernando Alonso had bumped into each other at dinner that night. Alonso had just about calmed down from the silent rage he had displayed in the paddock earlier on, but he was concerned that the delay to the stewards' decision suggested they were looking to exonerate Michael, rather than penalize him. If that happened, Alonso declared he was 'going to pull up on the grid, get out of my car and lie down on the grid in front of his.'

At Ferrari the famous loyalty and solidarity that Michael had worked so hard to foster was holding firm against the weight of evidence and the waiting reporters. In our interview after qualifying, I had put it to technical director Ross Brawn that the majority viewpoint in the pit lane was that Michael had parked his car deliberately to ruin Alonso's lap. Slightly mischievously, given Schumacher's track record, I asked whether Ross thought Michael would do that. 'Not the Michael I know,' replied Brawn, giving himself a sliver of wriggle room just in case the stewards found otherwise. As we waited into the evening, Ferrari's director of communications Luca Colajanni held court on the steps of the team's motorhome. Colajanni spoke for the Ferrari team and its boss Jean Todt, so while he, too, had to insist the incident was simply an accident, he was more interested in defending Ferrari's interests than just Michael's. Colajanni was in his element, picking apart arguments put forward by rival teams and drivers, getting into debates over their meaning, highlighting contradictions and hypocrisies and using his admirable grasp of semantics to re-frame how the sporting regulations may or may not have been broken at that moment at Rascasse. Inside the team motorhome Michael and Sabine Kehm knew they had a potential crisis to manage.

With most of his rivals safely in bed by this point, Schumacher appeared just after 10pm, with Jean Todt behind him, on the motorhome steps. We all held our microphones up as best we could (with the resourcefulness award going to the German-speaking crews who taped theirs together on to a boom pole and got a strong-armed sound engineer to hold them all directly under Michael's chin) and fired off questions. This was Schumacher's chance to have his say and to counter what he saw as overly harsh and personal attacks on his character, particularly those coming from 1982 world champion Keke Rosberg, who called Michael's manoeuvre 'the cheapest, dirtiest thing I have ever seen in F1'.

Schumacher had insisted in the press conference immediately after qualifying that it was a simple driving mistake. 'I was pushing, locked up and ran out of road, which is the consequence here in Monte Carlo,' he said. 'I didn't know what [lap times] the other guys were doing.' He added: 'You have your enemies and the people who believe in you. I won't be able to convince people. I've been here for many years, people should know better who I am and what I am.'

Maybe it was the lateness of the hour, but to me at least, Michael looked vulnerable. It wasn't like when he won the 1994 world championship in Adelaide by colliding with Damon Hill. On that occasion he had just realized his life's dream, and notwithstanding racing incidents along the way was on top of the world. It didn't even feel like the aftermath of Jerez in 1997 when, while eventually regretful, Michael was a little bemused at all the outcry, and even criticized rival Jacques Villeneuve for being too 'optimistic' in his overtake. This time, he was a third of the way through a season and was contemplating retirement, while his team was openly courting Kimi Räikkönen as a potential replacement. Nevertheless, he still had a realistic shot at an eighth drivers' title. The last thing Michael needed was everyone bringing up his propensity to go for an unfair advantage when the pressure was on.

When the stewards' verdict finally came, it was damning. Having analysed the speed at which Schumacher went into the corner, his 'excessive' braking and 'erratic' steering, and the TV footage seen by everyone, they effectively concluded that in 'deliberately stopping his car on the circuit' rather than, as he put it, just locking the brakes and running out of road, Michael had cheated. Even though the penalty of disqualification from qualifying (while still being allowed to start the race from the back of the grid) was seen by many in the paddock as lenient given the transgression, the reputational damage was significant. This was yet another moment when this ruthless racer had crossed the line.

On Sunday morning the paddock was full of jokes about Michael being the only Ferrari driver able to find a parking spot in Monaco. The man himself stayed tight-lipped for the rest of race day. As the last opportunity to ask him questions had been before the stewards' verdict, I wanted to get something on record from his camp. Thus just 12 hours after I had left the previous night, I made my way on Sunday morning to the Ferrari motorhome. After a long wait Michael's long-time manager and mentor Willi Weber came out, adjusting his sunglasses against the morning sun. I didn't have to beg for an interview, though – Willi was keen to talk.

'You know Michael, he's not an emotional man, it takes a little while before he reacts so I cannot tell you anything about his reaction,' he said. 'I can tell you about my reaction: I'm pissed!' It was classic Weber. Michael might have had a media image of being somewhat of a machine, but he was very much an emotional man underneath. By Sunday morning Weber would have known well his friend's reaction, and I assumed this was his way of letting us know Michael's state of mind. Weber went on to deny that the stewards' judgement of Schumacher's incident and the paddock's near universal condemnation of his tactics would have any bearing on Michael's decision on whether or not he would retire from F1 at the end of the year.

It was somewhat of a relief to finally get on with the race. If Schumacher was tired from Saturday's late night, it didn't show. Without help from any outside factors (a safety-car period actually hindered his race), Michael delivered one of the finest drives of his career to finish fifth from his pit lane start. He proved that you can indeed overtake in Monaco, and furthermore that he had no need to have performed the questionable parking manoeuvre at Rascasse. Even if he had started behind Alonso, his pace and that of the Ferrari and its Bridgestone tyres demonstrated he likely would have won the race. If he had done so he would have equalled Ayrton Senna's record of six Monaco victories.

Looking back on it with nearly 20 years of hindsight it's impossible to disagree with the judgement of the stewards, and the view that Michael used a racer's trick to gain a competitive advantage. In the 2020 Sky Documentaries programme *Race to Perfection*, Michael's 2006 teammate Felipe Massa revealed how 'causing a yellow flag' had been joked about in their pre-qualifying briefing, and how eventually, Schumacher came clean. 'It took one year for him to tell me he did it on purpose. One year. I said, "How can you do that?" It shows everyone makes mistakes in life.'

In the same documentary Ross Brawn gave this insight that forms a key part of understanding Michael Schumacher. 'Michael had occasional aberrations, things you could never give a logical explanation for,' Brawn said. 'He had this incredible competitiveness that drove him, and sometimes it would short circuit. Monaco pole, it's normally a given that you want it. But on that occasion, with the strategies, tyres and car we had, there was actually no need for it. It was just a stupid move. And one of those little glitches, short circuits that Michael had two or three times in his career.'

That kind of strike rate compares favourably with the very top drivers that came before Michael. He was the natural successor to Ayrton Senna and shared the great Brazilian's commitment to do anything to win. I always felt that Schumacher's job was harder, though, in that he was bridging two generations. He learned from the tough, uncompromising tactics of Senna, Prost, Piquet and Mansell, but also raced at a time when principles were changing. Drivers who followed him like Alonso, Räikkönen, Button, Massa and Hamilton were of the generation who believed the way you won, with fairness, honour and respect, was just as important as whether you won at all.

For all that motorsport had turned Schumacher into a hard and ruthless racer, off track he was not a hard man. He was kind, sensitive and thoughtful. It was his motorsport life that gave him his tough edge. His parents Rolf and Elisabeth were from a working-class background, and had taken on considerable debt to finance Michael's motor racing, so he felt huge pressure to succeed. As a teenager he would often travel to races by himself and as a result developed a high degree of self-reliance.

With the help of Willi Weber and Mercedes, Michael made it to F1, but nothing he experienced in his early races taught him this was anything but a brutally uncompromising environment. In 1991 Weber famously had to exaggerate Michael's knowledge of the Spa circuit when Eddie Jordan asked if Schumacher knew the track. 'Oh yes, he's driven it many times,' assured Weber, not specifying to Jordan that Michael had never driven Spa in a car, only toured round it on a bicycle. Schumacher and Weber shared a youth hostel room on his debut F1 weekend, with a toilet and basin between their beds – a long way from the glamour they might have expected. But Michael's talent and dedication made things happen for him. A tug of war for his services saw Mercedes motorsport manager Jochen Neerpasch and Weber pull Schumacher out of Jordan and into Briatore's Benetton team after that debut race. There he was afforded a greater opportunity to win, but also exposed to a frosty reception from three-time world champion Nelson Piquet, furious that Schumacher had replaced his friend Roberto Moreno. Tough business deals, tough teammates, tough racing – Michael was learning this was the way things were in F1. When I asked him about some of his strong racing tactics Michael would often talk about not wanting to 'give presents' to rivals. That stuck with me, helping me understand not only Schumacher, but many of the great drivers who came after him who have shared his uncompromising mindset.

His work ethic and his ultra-competitive approach were still there towards the end of 2006, but his mental batteries were running low. At the beginning of the season, he had started to think about how long he wanted to continue. That year's Ferrari was a good car with which he felt that he could challenge for an eighth world title, but internally, the dream team was starting to unravel. Throughout the year Ferrari President Luca di Montezemolo had made little secret of his admiration of Kimi Räikkönen's speed and wanted the Finn at Ferrari. Team principal Jean Todt preferred stability and wanted Schumacher to continue alongside Massa or possibly even Valentino Rossi, the MotoGP world champion who had been testing for Ferrari in a potential move from two wheels to four. Deals were being discussed, and there were plenty of leaks to us reporters from both sides throughout the summer. But by the time we arrived at Monza for the Italian GP the word was that Kimi was going to be a Ferrari driver in 2007. The question still open was whether Schumacher or Massa would be his teammate.

The Monza paddock is always the best of the season, the late-summer sun superheating every rumour running between the motorhomes. By race day, it was clear that Michael had been forced into a decision. On the grid, Martin Brundle asked Ross Brawn whether Schumacher really was going to retire from F1. 'You'll find out after the race, Martin,' which was as much of a 'yes' as you would ever get from Brawn. Schumacher made his point in the Grand Prix by outpacing Räikkönen to win for Ferrari, to the delight of their home fans. After the chequered flag I headed over to the Ferrari pit wall to grab whatever interview I could when I was tipped off that Ferrari were confirming Schumacher's retirement. I raced through to the paddock to find that copies of a press release were being handed out in front of the Ferrari motorhome.

Sheet of A4 in hand, I ran back to the pit lane to catch the end of the podium ceremony where I was able to drop a report into the beginning of ITV's post-race coverage. I was mid-flow confirming Schumacher's retirement when his engineer Chris Dyer stopped to have a look at the document. It was too good an opportunity not to get him to comment. 'Well, it's been a fantastic time with him, and we'll be disappointed, but it's his decision and hopefully we can finish the job this year and send him out in the perfect way,' he said.

Unfortunately for Dyer and Schumacher, the grand send-off wasn't to be. Later that day, Räikkönen was announced as a Ferrari driver for 2007. A month later in Japan, Ferrari suffered their first in-race engine failure for five years, effectively ending Schumacher's bid for an eighth world championship. All Fernando Alonso needed to do at the last race in Brazil to win was finish. With the pressure off, Schumacher headed into his final race weekend as a Ferrari driver in a pretty relaxed mood.

One thing he insisted on at Interlagos made me think back to our encounters over the years: no interviews. He wanted to enjoy the weekend and concentrate on the slim chance he had of beating Alonso to the title (to achieve that he needed to win, with Fernando not finishing at all), so we were told he would only pre-record a few quotes that would be distributed to all broadcasters, and there would be no one-to-one interviews. Someone in our ITV Sport production team, however, had a bright idea. Like every F1 driver, Michael liked receiving awards, so if we were to give him a leaving gift (so the thinking went), he could hardly turn it down. And if he agreed to accept it on camera, we would be able to get our own interview with him on his retirement that way.

The idea itself, while a bit sneaky, wasn't terrible. Alas what points it gained for concept were unfortunately lost in the execution. Our producer bought Germany's most famous sportsman, the football-loving Michael Schumacher, a replica England shirt from the 1966 World Cup Final, signed by Geoff Hurst and the remaining members of the squad that won that match 4-2 over . . . West Germany. Seeing as he was Schumacher's Benetton teammate back in 1992, Martin Brundle was given the job of presenting Schumacher with the large, framed, cross-of-St George-emblazoned shirt. Martin was embarrassed enough over the choice of gift that he stressed repeatedly to Michael in the handover ceremony that it was given 'in good spirit'! Thankfully Michael accepted it the same way. The only way it could have been worse was if we'd given him a 'no parking' sign, but he was generous enough to go along with it and gave Martin a few thoughts about his F1 retirement.

Alonso duly finished the Brazilian GP in second position to become a double world champion, while Michael saw the chequered flag in his final outing for Ferrari, coming home in fourth place. His father Rolf was at the race, and over the weekend we learned that Michael's seven-year-old son Mick had recently started racing go-karts. As he left Brazil to begin his retirement, one thing seemed likely, that Michael Schumacher was not quite done with motorsport just yet.

Off-track travel

Logistics, tips and tricks

Chapter 9

Nuts, Bolts, Bags and Flights

One of the many lessons I learned from Michael Schumacher was that 'you take your journey with you', and in F1 this is especially true. With 24 Grands Prix on the calendar from March to December, and with most of them squeezed into double and triple headers that require you to go straight from one country to the next, a trouble-free travel experience can set you up perfectly for the working weekend – just as delays, cancellations and disruptions can ruin it before a car has even turned a wheel. Over a typical season I spend around 370 hours on airplanes. Mercifully, budgets allow a certain amount of our air travel to be spent in the more comfortable seats on the plane – flights over six hours will be spent in premium economy or business class. A luxury, but one that does ensure I arrive at circuits reasonably well rested and ready to do my job without the need for a day to recover from a flight with no sleep.

For the F1 drivers and team principals, in recent years air travel has become even more luxurious, with many more journeys taken by private jet. The cost of these has become easier to reconcile

— 117 —

because travel is not covered by the FIA cost cap limits placed on all the teams, so any excess sponsorship or prize money that can't be used to develop the car can be used on flying privately. Max Verstappen has exclusive use of his plane while other drivers and team bosses lease time or are sponsored by business jet companies. It's not an environmentally friendly way to fly but does allow the top brass to make the most of their time at the circuits, or back home at the factory.

Before the cost cap, it was much more common to find yourself rubbing shoulders with F1 notables in the business-class cabin. It might be Sir Jackie Stewart, on his way to seat 1A, which was always reserved for him. Or imagine lying down for the night with Ross Brawn sound asleep not two feet away. Travel can lead to a strange intimacy with everyone enclosed in a pressurized tube at 39,000 feet – but the unsaid rule was always that work conversations stayed on the ground. Family holidays or the weather or the in-flight movie choices were all fine subjects to chat about, but the moment the plane touched down all that was forgotten as we got on with our jobs, hierarchy restored.

When British Airways still flew Boeing 747s, I remember dense fog once prevented us from landing in São Paulo and we were diverted to the city's third airport, Campinas. A moment of solidarity ensued as 30 F1 personnel all plotted furiously to figure out how to get from Campinas to downtown São Paolo – but luckily the fog lifted, and we were able to make the short hop back to the main airport, Guarulhos. There was also the Icelandic volcanic ash cloud of 2010 that grounded flights into and out of Northern Europe and left us all stranded in China, leading to a 'Great Race' style challenge for everyone to get back to the UK.

It was won, incidentally, by Australian driver Mark Webber, who had the notion to go via Dubai, and from there to Nice, hopping on a train that took him the last leg into London. As the ash cloud was moving eastwards from Iceland over Europe I decided to try to outrun it, flying east over the Pacific, but then had to wait it out in New York for a week before eventually making it home via Lisbon. Lee McKenzie, Jake Humphrey and the rest of the BBC F1 production crew flew to Frankfurt and then drove home via Calais.

Even under ordinary travel conditions, F1 travel is far from ordinary. Teams and media make extensive use of charter flights – where the travel company hires out the plane – both in Europe, and for some flyaway trips that are not served by scheduled flights, such as the marathon Las Vegas to Doha, Qatar, route. At Sky we also often travel on these charters with F1 team personnel, the main benefit being that the departure time is set to allow everyone to make it to the airport after the race. I've found myself on Boeing 737s operated by the Dutch airline Transavia, the TUI Fly Belgium or Poland's Enter Air, and Airbus A320s from the UK's Titan Airways. They typically take 180 people back from European cities to London's Luton Airport, which is equidistant from McLaren in Surrey to the rest of the UK teams based in Oxfordshire and around Silverstone. I'm fond of Luton, as it's only a 20-minute drive from my parents-in-law's house, which has proved a handy pit stop when things have been running late or I've had an incredibly early flight.

Even with a chartered flight, though, you can't be sure the journey will go smoothly. The first visit to the Russian Grand Prix in 2014 went without a hitch until it came to the Sunday evening return flight from Sochi to Luton. We had been scheduled to take off at 9pm, but the local aviation authorities unexpectedly cancelled every foreign operator's pre-booked take-off slot before re-opening them for sale.

While negotioations took place, the rest of us could do nothing except wait it out. I remember young British driver Max Chilton sitting on the departure-gate floor, leafing through reams of data from the day's race with his engineers. TV producers stood around, gossiping about what they thought of the race. Others perused the duty-free shops, remembering that they'd promised to pick up a set of nesting Matryoshka dolls for their kids. The team bosses were the first to get away in their private jets, while our charter dropped to the back of the queue. We finally made it home at 4am.

If you happen to be at an airport either side of a race weekend and you spot a group of 40–50 people dressed identically, you can be pretty sure that they are part of an F1 team. All teams work closely with clothing partners, and typical travel kit is smart jeans or chinos with crease-proof polo shirts or jumpers, embroidered with a subtle team logo just about visible to the eagle-eyed fellow flyer. Ferrari travel in smart blazers with a prancing horse logo, making them hard to miss. The ostensible reason for travel kit is to ensure that team staff look professional while representing their employer in public. A secondary benefit in years past was that if staff were more easily identifiable it would discourage any extended periods spent in airport lounge bars that could lead to unwanted incidents on long-haul flights (which, without naming any names, did happen from time to time).

One challenge with F1 team travel clothes is where and how to change out of them. When mechanics and engineers arrive at circuits they are expected to walk into the paddock in their race-team kit, so if they've had to drive straight from the airport to the track on a Thursday morning the only option is to change out of their travel kit in the airport car park. Many is the time I've walked past a load of mechanics in their underpants as they pack away their travel clothes and pop on the race-team kit, with full logos and branding, before getting into their minibus. The same is true at the other end of the weekend when team members must change back into their travel gear in the circuit car park before heading to the airport. If it has been raining this leads to all of them flying home in wet socks – truly the glamour of F1.

Luggage is also branded, usually provided by each team's official luggage partners. Fifty identical suitcases arriving at a baggage carousel would create chaos while people struggle to identify their bag, so they are all marked with a number allocated to each team member. Typically, a team principal gets 001, the drivers use their race number, the technical director tends to snap up 007 before anyone else can, and so on, in vague order of seniority. At one team this system caused such an argument among engineers that the pragmatic team boss ruled that bag numbers would be allocated at random to avoid resentment over who got the lowest numbers.

Having arrived, the next priority is somewhere to stay. Just as with the World Cup or Olympic Games, before a venue is selected to host a Formula 1 race a technical assessment is carried out, one element of which is ensuring there are enough hotel rooms within a reasonable distance from the circuit for the travelling personnel, VIP guests and attending fans. For races in or close to major cities such as Melbourne, Shanghai, Budapest, Singapore, Las Vegas or Barcelona it's rarely a problem, although thanks to the success of

Drive to Survive in boosting fan attendance in places like Austin and Montreal accommodation is increasingly hard to come by. At the more rural track locations, notably Silverstone, Spielberg and Spa, finding somewhere to stay within easy reach of the circuit can be challenging. For the first few years that F1 returned to Austria, we couldn't find anywhere to stay less than 45 minutes' drive from the track, so our production team were split between some lovely guest houses in the beautiful town of Murau, which was located next to the river Mur and had its own brewery. It felt like a true home from home as the charming owner of our *Gasthof* cooked us something delicious every night, although every dish included asparagus, for reasons unknown.

In recent years many F1 drivers have made their lives easier by staying in huge and luxurious motorhomes at the European races. They don't have to commute to and from hotels, and thus avoid traffic, and they can stay later at the track for evening briefings with engineers. The drivers hire people to maintain these giant RVs and to drive them from track to track, staying in a shared compound. Sometimes this is within the confines of the circuit, or a scooter ride away if there's no space at the venue, but what's always true is that for a few days it is probably the world's most exclusive campsite, with ultra-tight security on the gate to ensure that they are not beleaguered by fans.

Surprisingly tricky in terms of accommodation was the Korean GP, which was held between 2010 and 2013 before it was removed from the calendar for financial reasons. You might think that Seoul would be the ideal location for an F1 race, and Bernie Ecclestone had been keen for the Grand Prix to be hosted there. However, the group financing the race had their own preference, insisting on the

Yeongam district, close to the shipbuilding port town of Mokpo in the south-west of the country – a train journey or five-hour car ride from the capital. The brand-new circuit itself was fantastic, the facilities were top class, and despite the fact that it was somewhat out of the way, everyone enjoyed going there. Drivers, senior team personnel and the F1 and FIA top brass stayed in the nearby Hyundai Hotel, recently built on a hill with a fine view of the circuit and the neighbouring shipyard owned by the Korean conglomerate.

For the rest of us the only place to stay within an hour of the circuit was downtown Mokpo. There weren't many hotels there, or at least the kind of hotels that we were used to. What there were plenty of, as in many towns throughout South Korea, Japan and China, were establishments known colloquially as love hotels. In many cultures it's unusual for couples to live together before marriage, and while they're still residing in their family homes, it's difficult for them to get some privacy. That's where love hotels come in. The garishly decorated rooms are usually booked by the hour, so when 20 of us TV folk checked in, the receptionist (seated discreetly behind a screen) struggled to process the fact that we would be staying for five whole days and nights. Comparing the garish decor of the rooms (mirrored ceilings, theatrical lighting, no windows) and strange facilities (multiple computers and monitors in each room), the peculiarities of our respective love hotels were the main topic of conversation in the Korean International Circuit paddock. And were the rooms sub-let to young couples while we were at the circuit all day? We had our suspicions.

Assuming the F1 teams have their own hotels sorted, the first people at a circuit on race week are the setup crews, who arrive the weekend before the Grand Prix. Usually a team of five or six, sometimes including local contractors, they are responsible for preparing the team garages ahead of the race weekend. Each team has multiple identical sets of non-performance critical equipment, such as garage walls, tyre trolleys and plastic chairs which travel by truck in Europe and by sea freight for the flyaway events. These would have been delivered some time before the race and have been stored ready for the setup crews' arrival.

The garage framework has to be bolted together before the overhead gantries are installed that provide the power, ventilation systems and air lines. The restricted areas in the back of the garage are set up for carbon-fibre fabrication, gearbox servicing and engine preparation, all secret technology that needs to be hidden away from prying eyes. In every garage there's a little pod for the fuel companies to do their petrol and oil analysis, too. There's also a complex network of cables linking the pit wall to the garage, looping in the engineering trucks where meetings are held, and all of it connecting via fibre-optic network back to mission control at the team's factory, from where engineers and strategists are in constant contact with their colleagues at the track. Connections to the FIA data system and F1 TV are also plumbed in.

Teams always want a perfect garage floor – as level as possible, so the cars can be laser scanned accurately, and the smallest nut or bolt can be spotted and won't roll away when a mechanic needs it most. Any imperfections are filled in by the setup crew, and the garage floor is freshly painted at the beginning of the week. It's not just because teams want to look smart – it's so that any fluid leaks will show up easily, or any stray nuts or fasteners can be identified after a rebuild. The surface must also be non-slip but smooth enough to sticker with the team logo and the drivers' names and numbers.

Next to be constructed are the hospitality buildings used at European races, usually on the Tuesday before the race. We used to call them motorhomes, because that's exactly what they were, but these days they are elaborate structures that travel on 20 or more trucks and require a crane and crew to erect them and take them down. Each team's building is bespoke and has its own level of complexity. While Ferrari, for example, has a sleek Italian elegance with darkened windows and the iconic red trim, Aston Martin incorporates fine lines and classic leather and wood in the interiors. Once they have been constructed, they are ready for the hospitality staff to start preparing for the hundreds of team members and guests that they will serve over the weekend.

Next in are the mechanics, who will typically arrive on Tuesday, taking over from the setup crew, and start to prepare the two race cars on their perfect garage floor. Engineers arrive on Wednesday afternoon, while drivers will start appearing on Thursday lunchtime ready for their media and briefing commitments. The last members of the F1 race teams to arrive are usually the team bosses. With only a couple of exceptions (Ferrari's Fred Vasseur, McLaren's Andrea Stella), team principals don't feel the need to be at the circuit on the Thursday before track action starts, generally showing up on Friday morning having had a precious extra day in their office back at the factory.

Qualifying and race days come and go, and before the champagne is even dry on their race suits the drivers are whisked off by helicopter or driven under a police escort back to their private jets at the local airport. For the rest of the team members and the media the Sunday evening dash to make flights is often the most stressful part of the weekend, especially after a long and tiring race day. That's why it can sometimes be comforting to know that you're on a charter flight that at the very least shouldn't leave without you if the race has been delayed. Mechanics, meanwhile, usually spend Sunday evening working on the cars and preparing everything for its journey to the next GP, be it by air or road. They will usually fly home, or to the next venue, on Monday. In the case of triple-header races, therefore, it's likely that F1 mechanics won't see their families at home for 22 days straight.

If you're considering going to watch a race, which one should you go to? It's a question I'm often asked. In Europe, number one has to be Monaco. One of the oldest events, and the most prestigious, the sport's most famous street race never disappoints. The late spring weather is usually beautiful, the history and heritage are alluring, and the general glamour of Monaco is intoxicating, which is just as well when you find out how much it costs to buy a drink. Nowhere else allows you to get closer to the cars and the action than Monaco. During the build-up to the weekend, you'll frequently spot drivers, team bosses and other F1 luminaries as you walk around the principality. Monaco is also the only circuit that opens fully for the public to walk around every evening, even through the tunnel. It's no small thrill to stand on the start/finish line or place yourself on the pole-position grid box just a few hours before the Grand Prix starts. It might not be the cheapest race in terms of ticket prices, but being there is a pretty magical experience – and the tip for those on a budget is to stay in Nice and commute in by train.

Otherwise, the festival-like atmosphere of Silverstone is a lot of fun. Austria is always extremely well-organized amid breathtaking scenery. Budapest is a beautiful city to visit in the middle of summer, and there's plenty to do in town for any family members less interested than you in F1. Baku's old town is worth discovering if you can find space on a flight to Azerbaijan. And then there's Monza – a late-summer gem of a Grand Prix, where the historic atmosphere is almost palpable, hanging between the trees and the old banking. Spa is impressive due to its sheer size – but bring your raincoat! Logistics are always a major consideration, of course. For many European races camping at or close to the track is a good budget option, and you can combine a race trip with a holiday.

Of the long-haul races if travelling from Europe, those I always recommend are Melbourne, Suzuka, Montreal, Austin, Singapore and my personal favourite, Interlagos in São Paulo, Brazil. Again, the city races are obviously easier logistically, and while Japan and Brazil have their challenges in terms of getting around, they are both fascinating places.

If you are travelling to a F1 race, here are some of my essential tricks of the trade to save time and money. Most importantly, as soon as the F1 calendar comes out, get booking! Most people wait, but you can snap up a bargain on flights and hotels before the operators realize there's going to be a Grand Prix in town. Sign up to the airline's frequent flyer programmes and their email marketing lists – they often have seat sales that you can use to fly to far-flung F1 destinations. It's the same for hotels in the destination city. Pack light – the key to saving time in airports is not to check in any hold luggage. That allows you to check in online, and head straight through security, and you'll also save time at the other end at baggage reclaim. If you're going away for a week or less, the only thing that's stopping you taking hand luggage is self-disciplined packing. If you're worried about your carry-on not being allowed in the cabin, invest in a soft bag that you will be able to squeeze into the annoying gauges that they have at the gate.

Public transport is always the best and often the quickest way of getting to and from airports and circuits, but if that's not possible then ride-share apps would be my preference for ensuring you won't get ripped off. Take a look at which side of a city the track is located, because there are often better options to stay in other areas. For the Miami GP, for example, it's more convenient to stay in Fort Lauderdale than Miami itself, while San Antonio is an affordable option for Austin. The city of Girona is within reach of the Barcelona circuit, as are many Catalan coastal resorts, while for the Dutch GP the town of Haarlem or Amsterdam itself are much better options than the small town of Zandvoort.

One last tip that I have never been able to do myself is to go with the flow and join in with the locals. No matter who your favourite team or driver, at the Dutch GP get yourself some Max Verstappen merchandise and hang out with the Orange Army for a weekend. Alternatively, feel the Ferrari passion, find some red trousers, and join the *tifosi* at Monza. You'll never be short of a friend to share a pizza with, and you'll have the time of your life. See you trackside!

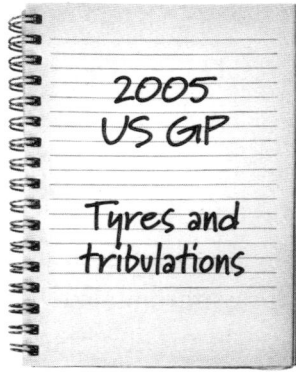

2005
US GP

Tyres and
tribulations

Chapter 10

The Six-Car Race

Reporting is easy. You come on, say your piece and then stop. Presenting the show is much harder. There's a good reason they call the presenter of a news or sport programme the 'anchor'. A live programme is like a swirling body of water, and there are strong currents that sometimes move things from where they're supposed to be. The anchor keeps everything stable. Or, to use more of a cheese-based metaphor (my preferred foodstuff for this sort of thing), being a successful presenter is like being able to take many strands of stringy cheese, process them together in your brain (while talking) and deliver them to the viewer as a perfectly shaped, easily understandable ball of mozzarella.

How do I know presenting is hard? I've tried it. Presenters need to have an incredibly good short-term memory so that they can remember their next link while listening to whatever is going out on-air and the instructions they're being given by the producer. Throughout my years at Sky Simon Lazenby has anchored our programme. He came to F1 from the broadcaster's live Rugby

Union coverage via a spell on a trading desk and running 'The Chilli Shack', a street food stall housed in an old fire truck. Simon is so good at what he does, he sometimes memorizes not only all his links (we don't use autocue) but also the whole running order and then hands his print-out to one of the camera crew. You know you're watching a master at work when you see Simon presenting one of our programmes holding just a microphone (and without a rolled-up running order in his back pocket). 'Not bad for a links man' is the general comment that I and our regular camera crew of Pete, Lee, JD and Keiran routinely throw Simon's way after he's successfully piloted a complex programme to its conclusion, sometimes accompanied by a slow hand clap if he's done really well. If he knew how full of admiration we are for his trademark smooth delivery, on-point questions and quick wit, we'd never hear the end of it.

Over the years I've worked with many different show anchors. During my time at the BBC, I saw Jake Humphrey move from presenting kids' TV to being one of the most sought-after sports broadcasters in the country. Outside of some experience working on football, presenting Formula 1 on the BBC was the first major anchor role for Humphrey, and he brought a wide-eyed enthusiasm that perfectly countered the 'been there, done that' experience of David Coulthard and Eddie Jordan.

A Norwich lad, Jake was following in the footsteps of Steve Rider, who had performed the F1 presenter role at ITV. Rider's career path had gone from reporting from Norfolk's Snetterton circuit on the Formula 3 battles between Martin Brundle and Ayrton Senna, via the Olympic Games, several World Cups, European Championships, Commonwealth Games, Masters golf and many seasons of *Grandstand* to anchoring the last two years of Formula 1 coverage on ITV. Steve Rider presented F1 throughout my formative years when it was screened as part of *Sunday Grandstand* and his reporting from Imola in 1994 after the deaths of Roland Ratzenberger and Ayrton Senna had a huge impact on me. Rider retired midway through 2025 and, for me at least, was the doyen of sports broadcasting. As someone who had worked in print journalism before coming to television, his scripts and ability to find the perfect words were second to none, always matched by his flawless delivery.

But there was one race in particular that showcased better than any other the ability required of an F1 presenter to rip up the running order and wing it, and that was the 2005 US Grand Prix. The man in the chair for what became known as 'the six-car race' was Jim Rosenthal.

It had already been an interesting season. After winning five straight world championships, Michael Schumacher and Ferrari had been tripped up by a single rule change that left the Italian team struggling. This was the era of an intense tyre war between the two tyre suppliers Bridgestone and Michelin. Much of Ferrari's success over the previous seasons had been down to its close relationship with Bridgestone, whereby, as the driver most likely to win races and championships, Schumacher had been able to develop the tyres to his personal preference. Gradually the other big teams such as McLaren, Williams, Renault and Toyota all moved to Michelin, not wanting to be tied to Ferrari's tyre-development direction. The move to Michelin brought them a reasonable amount of success, but the Ferrari–Bridgestone bond was so strong that from 2000 to 2004 Michael won every drivers' championship.

In 2005, in a move that would shake the championship up, the FIA introduced a dramatic rule change – drivers were required to use only one set of tyres for the whole race, and pit stops would be for fuel only. Michelin did a much better job of adapting to this new challenge than Bridgestone, who had previously been so good at making the 'sprint' tyres that had allowed Schumacher to run flat-out over short stints. The only major team affected was Ferrari, as the other remaining Bridgestone customers were backmarkers Jordan and Minardi. With Michelin suddenly dominant, the 2005 season quickly developed into a fight between Fernando Alonso of Renault and McLaren's Kimi Räikkönen. When we arrived at Indianapolis in June Schumacher was running fifth in the world championship, and Ferrari hadn't yet won a single race.

The weekend started like any other, but the first sign of trouble came in the first practice session when Toyota's test driver Ricardo Zonta had a spin into the gravel trap. Replays showed that he'd had a left-rear tyre failure, but that could have been caused by debris or any number of things, so there wasn't too much concern. What really got everyone's attention was when, in second practice, Zonta's teammate Ralf Schumacher had a high-speed crash at Turn 13. Indy's banked and extremely fast final corner had been a focus of attention since F1's first race at the venue five years earlier, but this time the TV footage indicated that he too had suffered a left-rear Michelin tyre failure just like Zonta's in the morning session.

Behind the scenes, this was Michelin's worst nightmare. After Friday practice the French company's engineers examined the rear tyres from all their teams, and to their horror, discovered that many of them had flaws in the sidewall – defects that were weakening the structural integrity of the tyre. Michelin suspected that the sheer forces on the tyre on the banked Turn 13, combined with the way their tyres were made, were effectively rendering their product unreliable. Worse still, there was nothing that could be done to mitigate it.

On Saturday morning we had a production meeting. By then a few of us had heard from the teams that Michelin was concerned about the double tyre failures on the Toyotas. I remember interviewing Pierre Dupasquier, the charismatic Michelin competitions boss, and Nick Shorrock, who was the company's top engineer, but neither of them had let on that this was a problem that wasn't fixable. They were more upfront with the FIA and warned race director Charlie Whiting that their analysis suggested that the flaws seen in the rear tyres would likely lead to tyre failures after 15 laps at racing speed. The US Grand Prix at Indianapolis was a 73-lap race.

There were no easy solutions. The Michelin-shod cars would either need to change tyres four times in the race, which would look highly unusual and incur penalties for breaking the new 'single set of tyres' rule. Even if that rule was not enforced for one race due to *force majeure*, the consequence would be to create two separate races, between the one-stopping Bridgestone teams and the four-stopping Michelin teams. The only other option was to change the circuit – slowing the cars down into Turn 13 – which would reduce the loads that were damaging the tyres.

The F1 timetable doesn't stop for anything. While discussions continued and letters were going back and forth, we headed into qualifying. In those days you carried the fuel for your first race stint in qualifying, so qualifying as well as possible was a delicate balance, trying to use the absolute minimum of fuel for a light car and therefore a quick lap time while not compromising your race by only having enough fuel to do a few laps, forcing an early pit stop while others with more fuel on board overtook you.

We expected the qualifying battle to be between title contenders Fernando Alonso and Kimi Räikkönen. Instead Jarno Trulli put his Toyota on pole position. The immediate assumption in the pit lane was that Toyota was trying to save face after its two practice incidents by running Trulli light, winning him pole. What we didn't realize at the time was that Trulli's engineers had indeed run him very light on fuel because they knew that their cars would either need to pit after a few laps or, worse still, would not be competing in the Grand Prix at all.

Following some more consideration and analysis after qualifying, Michelin's management raised the stakes. They told the FIA that they would not allow their teams to race around the circuit as it was, even if they made four pit stops. If the FIA wanted there to be a Grand Prix, stated Michelin, they would have to change the circuit's layout to slow the cars' entry and exit speeds at Turn 13. The easiest way to do this would be to paint new lines on the track and install a few bollards in order to create a tight chicane through which the drivers would have to pass at reduced speed.

Asking the FIA to change the track layout and create a new chicane just because one tyre manufacturer had supplied deficient material was unprecedented and unfair, a point that Bridgestone and Ferrari made vigorously. Bridgestone tyres were fine through the fast Turn 13, the loads causing no problems. Clearly, Ferrari also saw this as an opportunity to score some valuable points during what had been a difficult season, and to get themselves back into the championship hunt. It's also fair to say that Ferrari boss Jean

Todt might have been feeling somewhat victimized by the sudden 'one race, one set of tyres' rule and wasn't in any kind of mood to give the Michelin teams a free pass.

FIA president Max Mosley wasn't at the race, but he was being kept closely informed by race director Charlie Whiting about events and the chicane discussions. Mosley had a brilliant mind, which came with a mischievous streak, and when I interviewed him about the Indianapolis saga at the next race in France, he referred to the chicane off-camera as 'the chicken'. I never found out why, but I suspect he was referring to the affected teams 'chickening out' by not racing. Nevertheless, the FIA were perfectly within their rights to reject the chicane idea. There are strict rules concerning the approval and licensing of circuits for safety reasons, and everything must be homologated – legally signed off – by the FIA before the race weekend. Another factor was that the cars hadn't been fitted with gears suitable for a sudden decrease in speed. They were geared to pick up speed through the last proper corner and then go flat-out around Turn 13 and down the main straight.

Whiting's suggestion in turn was that the Michelin teams should simply decrease speed around Turn 13 in order to protect their tyres. He was willing to allow a white line to be painted around the corner – the six Bridgestone cars would be allowed to run at normal pace above it, and the Michelin cars would run below it with their speeds monitored by a radar gun. There would be, in effect, two parallel races. This was, at least, a workable if not desirable solution, but everyone went to bed on Saturday evening not knowing what tomorrow would bring.

Sunday morning at Indianapolis dawned bright and breezy. The 100,000-strong crowd of spectators made their way into the circuit, past the coffee and donut stands, anticipating an exciting race. Fans had travelled from as far as Colombia to see their hero Juan Pablo Montoya compete, and from across America and Europe, to get a feel for Formula 1 at the home of the Indianapolis 500, the USA's most famous race.

Aware discussions were still going on to try and find a solution to the Michelin problem, we had a quick production meeting, before Louise Goodman and I went into the paddock and started interviewing anyone we could find, to find out what was really going on. This was the moment Jim Rosenthal really came into his own. As the presenter, he knew how important it was for him to hold the whole programme together, processing that stringy cheese into mozzarella. So, he made a point of listening to all the interviews that Louise and I were getting in the paddock throughout the morning, with team bosses like Eddie Jordan and Minardi's Paul Stoddart, both Bridgestone runners, and McLaren's Martin Whitmarsh, whose cars were on Michelin. I remember interviewing David Coulthard on his way to the drivers' parade, and Mark Webber on his way back from it. All Michelin drivers were quite clear that they wanted to race, but that the chicane was the only realistic solution.

Meanwhile Louise got the other point of view. Ferrari was only too ready to race around the circuit the way it was. The other Bridgestone runners, Jordan and Minardi, were more open than Ferrari to finding a compromise solution. Yet Ferrari's position was not unreasonable. 'This has nothing to do with us, we'll leave it to the FIA' Jean Todt stated, while knowing full well that the FIA weren't going to allow any significant changes to the circuit. Every potential outcome was going to be beneficial for Ferrari – either the

Michelin teams weren't going to race, or they would run their own race and inevitably finish outside the points. The Bridgestone teams had equipment that wasn't going to fail. Ross Brawn made a clear point about how it would be the same with any other kind of component on the car. A team might come to a race knowing that their brakes weren't going to work, or the engine wasn't going to last beyond 10 laps, and tyres were no different. The decision for the Michelin teams was either not to take part, or to drive until the tyres failed, and then retire from the race.

As the Grand Prix approached, the meetings became increasingly frantic. There wasn't time for secrecy anymore, the paddock offices had huge windows, and from the outside it was easy to see everything that was happening. It was extremely entertaining to watch the team bosses arguing with each other – Ron Dennis shaking his head, Flavio Briatore waving his arms around, the drivers just standing there, looking like spare parts in this whole discussion, bemused, not knowing what was going on.

Meanwhile, we had a whole running order full of pre-recorded features lined up for our race show on ITV, including one that I had done on how F1's popularity in the USA was really picking up. That was the first piece to get dropped. And then one by one more features bit the dust. The running order was literally ripped up, replaced by Jim Rosenthal and Mark Blundell's analysis, and live interviews from Louise Goodman, me and Martin Brundle coming in from the paddock, as we tried to follow the unfolding story and keep viewers up-to-date with the drama.

I remember very well Jim's last instructions to our director Simon Dukes and editor Gerard Lane – keep the communication clear, keep it simple, and keep telling me where we're going next.

And that's the presenter's skill again. They need to know what's coming up while listening to the interview being played in, in order to be able to react to it and make more sense of it for the viewer. The hard thing about ripping up the running order and following where things went was that we never really knew where the story was going to go.

The drivers' parade took place, and the frantic meetings in the paddock offices continued. And then as the opening of the pit lane approached the drivers got ready to get in their cars. The confusing element for us in the pit lane was that as a result of all those interviews we were pretty certain that the Michelin runners would not be able to race. All the plans had failed, all the options had been exhausted. It would be a six-car race with the Ferraris, Jordans and Minardis. But when the pit lane opened, all the cars went to the grid. What we didn't know (not having access to the teams' contracts with Formula 1 and the FIA) was that the Michelin drivers had been told that legally, they had to drive to the grid to fulfil their team's contractual obligation to 'participate' in the event. They would then have to come into the pits and retire at the end of the formation lap.

This was going to hurt. Being six hours behind UK time, we were in a peak TV slot in the UK on Sunday evening, ahead of long-running British television soap opera, *Coronation Street*, so we had to explain to one of our biggest audiences of the year that they would be watching a six-car race. Martin did a brilliant grid walk. By now he knew from sources that the plan was for the Michelin teams not to race, despite there being no announcement of any kind. He pulled no punches in explaining what was likely to happen. He gave Bernie Ecclestone a good grilling, asking him big-picture questions about the impact on F1 in America and the future of Michelin in the sport. Bernie replied flatly: 'Not good in both cases.' Martin ended by asking, 'Surely we just have to have a sensible pill and say, "OK, this is the situation, let's take a sensible solution and go motor racing."' 'Tell me where to buy the pills,' said Bernie.

Sadly, the sensible pill never appeared. We could hardly believe what was about to happen. We still had some hope that at the last minute something was going to change. After all, everyone was on the grid. Surely they wouldn't be doing that if they were just going to head back into the pits? Even on the formation lap David Coulthard came on the radio to his Red Bull engineers: 'If it comes down to the drivers, I want to race.'

The cars were halfway around the formation lap when I noticed McLaren mechanics folding up all the chairs they had carefully laid out in the garage. I called through to our editor Gerard Lane, 'I've got a line from down here!' James Allen said, 'We don't know what's going to happen. Ted, do you have any more information?' 'James, I can see Renault and McLaren, they're both clearing two F1 car-sized spaces in the garage for their drivers to come in and retire.' There was also a radio message to Fernando Alonso from his Renault engineer Rod Nelson, who said, 'OK mate, you know what the plan is for the start, straight into the pits please.' Perhaps that was intended to discourage the world championship leader from going rogue!

Led by pole-sitter Trulli, all the Michelin runners peeled into the pit lane, leaving the six Bridgestone drivers to go the grid – the two Ferraris of Michael Schumacher and Rubens Barrichello in fifth and seventh spots, and the Jordans and Minardis at the back, with a huge empty space between them. I remember watching the start of this six-car race, just staring at the monitor, not quite believing what I was seeing. In retrospect, everything that happened on Saturday night and Sunday morning had led inexorably to this point. The Michelin cars couldn't race without a chicane and a chicane couldn't be made.

The race was interesting for about a lap and a half, watching Schumacher get away ahead of Barrichello after an early tussle. On lap two, I realized that there wasn't going to be anything worth reporting on from the pits, so I headed into the paddock. Louise and I grabbed a few interviews with some of the Michelin drivers and team bosses expressing their disappointment with the outcome. And then it occurred to me that I should get out into the public areas to see what was going on there. It was the first time I'd done this in the middle of a race, but instinct told me that was the place I needed to be. Reassured by our tech people that my radio camera and microphone would work at that distance, off I went through the tunnel and out to the back of the grandstands.

There were many fans from around the world – and they were all confused. They didn't really know what was going on, or why there were only three teams racing. This was before the days of social media and internet on phones, so half my time was spent telling them the whole sorry story.

Some fans were shouting for refunds; others were starting to leave the grandstands. The bit I remember best was the strangest mid-race interview I'd ever done – with an Indianapolis police officer. He said, 'I don't know what's going on, but it's starting to get a bit sketchy down here. We've got a lot of angry people throwing their cups of beer around.' Implicit in his tone was 'What have you stupid Europeans done here?'

Meanwhile, Schumacher beat Barrichello to win the race. The only tension was over which Jordan driver was going to be on the podium, but even that wasn't particularly close, the likeable Portuguese Tiago Monteiro beating teammate Narain Karthikeyan, who himself became the first Indian driver ever to score points in F1. The Minardis finished fifth and sixth, an outcome that the team could usually only dream about. However, none of it meant much to the fans, who made their displeasure clear, and the boos could be heard from the paddock.

Towards the end of the post-race part of the programme Jim Rosenthal had to do a promo for the upcoming *Coronation Street* episode. '*Coronation Street* is coming right up for you, more complications for Steve and Tracy. That is a soap opera worth watching. We had a bit of a soap opera here as well . . .' Jim ended the programme by saying, 'You've seen an F1 fiasco in peak time, and like David Coulthard I feel sick and embarrassed to my stomach. Circumstances beyond our control. We can only say sorry. Good night.'

It was quite strong stuff from Jim, but he was keenly conscious of the feelings of the viewers. That's what the presenter is there for – they need to ask the questions that the viewer wants answered, and to walk the viewer through events as they unfolded. Nobody else had offered an apology. Jim thought the viewers deserved one.

We went away from that race with mixed feelings. Frustrated, bemused and pretty disappointed, yes, but also content with the fact that as professionals we had stayed on top of the story, heard from all the key players and had the information to explain what was going to happen and why. The US Grand Prix lasted two more years at Indianapolis before falling off the calendar, while Michelin withdrew from F1 at the end of 2006, with Bridgestone becoming the sole supplier until Pirelli took over in 2011. Michelin has not subsequently been interested in returning to Formula 1.

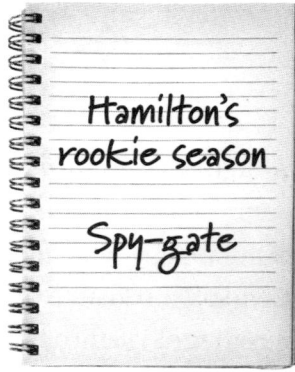

Hamilton's
rookie season

Spy-gate

Chapter 11

From Ron to Ruin

———

Sometimes in F1, like the parable of the frog in the boiling pot of water, it's easy not to notice that significant events are happening around you. So much happens every day of every race weekend, it can be hard to see beyond details that feel like the continuation of ongoing stories. However, the transfer of eras in F1 between 2006 and 2007 was the kind of immediate event that would have that frog hopping right out of the pot.

Michael Schumacher was gone – manoeuvred into retirement, as we'd later discover, by Ferrari president Luca di Montezemolo – and for a few months at least, the Fernando Alonso era was at its peak. At the end of 2006, Alonso was a double world champion, and for 2007 he joined McLaren, the team of his hero Ayrton Senna, where his wish and expectation were to match the Brazilian legend's tally of three world titles. Unfortunately for Alonso the man who would dominate the next era of F1, and who happened to be his new teammate, had other ideas.

Lewis Hamilton was a McLaren protégé and was still an F3 driver when Alonso first signed for McLaren in December 2005. Many observers of the junior categories had identified Hamilton as the next hot prospect and were keen to see what he did on his graduation to GP2, the next step up the ladder, in 2006. Possibly because of how promising Hamilton looked, and possibly as a favour to Bernie Ecclestone, who was keen to raise the profile and popularity of the GP2 series, ITV started covering it with a standalone programme, and asked me and Louise Goodman to present it.

Our GP2 coverage tended to be planned after we'd had our discussions about the F1 stories of the weekend. At the end of our Thursday morning meetings at European races with GP2 on the bill, we'd often be told that we had an interview with Hamilton in the support-race paddock. Generally Louise and I would take it in turns. We'd find a stack of tyres to sit on near his ART team's race truck and interview Lewis about how his season was going. The GP2 championship had quickly developed into a fight between him, a rookie, and the more experienced Nelson Piquet Jr. Hamilton came across as pretty shy back then, and even in those days clearly regarded interviews as a bit of a burden that he wished he didn't have to do. He was wary of saying the wrong thing with McLaren personnel listening, but he was also resigned to the fact that it was a necessary part of the job. On our side, sensing that, we tried to make interviews as enjoyable and efficient as possible, but it was when the cameras stopped recording that Lewis would perk up and grill either me or Louise for information about what was going on at McLaren, what we thought their thinking was on drivers and who else they were looking at alongside Alonso, as well as filling in his knowledge about the car's performance, any information that might help his career.

As that season went on, and Lewis started winning consistently, there was a sense of excitement whenever he was on the track. So much so that the GP2 races, often ignored in the F1 paddock, started to become essential viewing in team motorhomes – especially at McLaren. Ron Dennis and Mercedes motorsport chief Norbert Haug would sit downstairs in one of the glass-walled offices of their motorhome watching the races enthusiastically, to the obvious irritation of McLaren F1 drivers Kimi Räikkönen and Juan Pablo Montoya as they passed behind to get to their driver rooms.

In Monaco, Lewis appeared on the Grand Prix grid, having won that weekend's GP2 race in some style. Strange as it might seem today, Lewis didn't routinely get an F1 paddock pass, never mind grid access. He had been invited on as a result of his win, but he looked a bit self-conscious as he stood behind Räikkönen's car. Martin Brundle had just finished interviewing the Duchess of York, Sarah Ferguson, who had revealed that rather than either of the drivers she was 'a great supporter of Ron Dennis'. Martin then turned to find Lewis standing nearby.

Martin: 'Lewis, do you think you might be standing here next year on the grid?'

Lewis: 'Hopefully, it's a fantastic feeling to be here, I just hope I can sit there next year.'

Martin: 'OK – and you think you'll be fully ready for it?'

Lewis: 'I think so . . .'

Lewis's self-confidence was growing week by week. He'd had a breakthrough about a month earlier, when, following a tricky start to the season, he'd dominated both GP2 races at the Nürburgring. The Monaco win confirmed his form. There was a lovely moment just after Brundle's interview when Sir Stirling Moss came over to Lewis to say hello and congratulate him on his victory the day before. The two would get to know each other well over the coming years, with Moss often talking about how he admired Hamilton's racing skills.

As 2006 progressed it became less of a question of if Hamilton would graduate to F1, but more when, and with which team – specifically whether he was ready to be a full McLaren F1 driver. 'I don't expect to jump straight into a McLaren,' Lewis had said. 'I would have thought I would go into one of the lower teams, but we'll just have to wait and see.' The problem with that plan was McLaren didn't have any friendly teams lower down the grid that they could loan Lewis out to for a single season. As he kept winning in GP2, Ron Dennis and Martin Whitmarsh grew increasingly convinced that they had Alonso's teammate right there, ready to go. Although they didn't rush to tell Hamilton just yet. When I interviewed him at the start of the Turkish GP weekend in August, Lewis was clearly none the wiser, asking whether I thought McLaren were seriously looking at anyone else for Alonso's teammate, because as he was aware, they certainly had other options.

It was on the grid of the Italian GP two weeks later, after Hamilton had secured the GP2 title and his own racing season was over, that Dennis finally informed Lewis of his decision. 'I'm going to give you a chance,' he put it, simply. A year later in an ITV interview with Steve Rider, Lewis admitted that he hadn't been certain exactly what Ron meant, but he wanted to sound professional, so he just replied, 'OK, Ron, thanks very much'. What he hadn't fully realized was that Ron had decided that his protégé was indeed ready, and that Hamilton would be racing for McLaren in 2007. The team devised an extensive winter test programme for him, with way more mileage than young drivers can log these days under the current regulations.

From Fernando Alonso's point of view, Hamilton's appointment was something to take note of, but he didn't see it as a threat. My sense was he thought, 'Ah, a rookie alongside me, that's fine. He's not going to be a problem.' Alonso was at the height of his powers. He'd won two world championships, had signed a contract with McLaren for very much more money than he had been getting at Renault, and was set to be the star driver of his new team. What I also remember noticing at the time was how Alonso threw himself into what he thought was the McLaren way. He had his hair cut short, very neat and tidy, was clean-shaven every day, and unveiled a new crash helmet design which was predominantly silver and black. In effect he 'McLaren-ized' himself. Growing up, he had shared his father's admiration for Ayrton Senna and he'd always had a soft spot for the team. His dad had even built him a little pedal kart in the famous red and white Marlboro McLaren colours. Alonso made it clear how it was a dream come true for him to race for McLaren, and the team were equally enthusiastic about him in return.

One example of that was the choice of a Spanish venue for the team's car launch, the architectural marvel that is the City of Arts and Sciences in Valencia. McLaren's new title sponsor, Vodafone, was keen to make a splash, and money seemed to be no object.

British-based media representatives were told to report to Farnborough Airport that morning, and were ushered on to chartered jets to be flown to Valencia and back for the launch day. I sat with other F1 journalists admiring the 'Hemisfèric' planetarium and ooh-ing and aah-ing at the fireworks, while a series of acrobats, stilt walkers and break dancers entertained us all. Eventually out walked Alonso and Hamilton, and the new car was unveiled. It looked fantastic in its chrome and bright 'rocket' red livery and was, aerodynamically, very detailed. Everything was going swimmingly – Fernando spoke like someone who justifiably saw McLaren as his team to lead, and as reigning world champion he assumed that he would be afforded some level of seniority and preference in the way they went racing.

Alonso didn't have to wait long to find out that he wasn't going to get it. At the first race in Australia, Hamilton couldn't have announced his intentions more clearly, overtaking his more experienced teammate off the start line. Fernando did then pass him with pit-stop strategy, and they ultimately finished second and third, behind Kimi Räikkönen's Ferrari. However, it was obvious to everyone that Lewis had not come prepared to be a number two supporting act to Fernando. He was there to race and to win.

Alonso scored victories for McLaren in Malaysia and Monaco, while Räikkönen's Ferrari teammate Felipe Massa triumphed in Bahrain and Barcelona. But at every one of the first five races of the season, Hamilton was on the podium logging points. Lewis scored his maiden victory at round six in Canada, and followed up immediately with another in the USA. He took the lead of the world championship, while Alonso had a couple of poor results in Canada and France, and found himself lagging behind his young teammate and growing pretty annoyed at McLaren for not, as he saw it, helping him maximize his points. Hamilton was certainly aware of the situation, acknowledging later in the year, 'Nobody expected me to be as competitive as I am, leading the world championship, and having the current world champion chasing me.'

Any F1 newcomer who found themselves leading the championship in their rookie season, albeit driving the grid's fastest car, would have been a highly noteworthy story. As the UK's free-to-air broadcaster it was clear to us at ITV that we had a major new British sporting star performing every race week in our F1 coverage, someone our viewers were keen to know more about, and as a result we closely followed Hamilton's exploits.

After qualifying at most races, Steve Rider would do a sit-down interview with Lewis, which McLaren were keen to facilitate because it gave Vodafone and their other sponsors a generous amount of screen time. These days it's unimaginable that Lewis would do a 15-minute one-to-one interview with a broadcaster after each qualifying session, but as a rookie, he didn't know any different – the team wanted him to do it, therefore he obliged. And as we'd find out as the season went on, Rider's calm, reassuring manner allowed Lewis to feel comfortable enough that he revealed insightful details about what was going on at McLaren at the time.

We were of course equally focused on Fernando Alonso's role in the title fight, and could understand his conviction that he should be prioritized as team leader to contend for the drivers' title. However, his boss Ron Dennis had just as strong a resolve. Unlike Ferrari during the Schumacher era, one of the many principles Ron was committed to was that of equality between the two McLaren cars. Alonso appreciated Ron's position, but when he saw Hamilton starting to disobey team instructions, that was when things really started to fray. The most dramatic example was in qualifying for the 2007 Hungarian Grand Prix.

I was watching the session from a corner of the McLaren garage thanks to a longstanding deal I had with McLaren's then sporting director, Dave Ryan. A deal, you might wonder? Let me explain. In all my years in F1, I have found only one or two people more intimidating than Dave Ryan, a New Zealander who joined McLaren in the 1970s. One of my first tasks at ITV Sport in February 1997 was to connect our camera crew up with him at a Silverstone test in order that we could get some shots of David Coulthard for our programme titles. Familiar with Ryan from TV and the Ayrton Senna videos I'd grown up watching, I spotted him straight away, and enthusiastically approached as he emerged from the McLaren race truck. What followed can best be described as 'character-building' as Ryan proceeded to tear strips off me for wearing a North Face puffa jacket (which I had donned as a precaution against the famous Silverstone wind), rather than some (non-existent, I would take a second here to note) ITV Sport-branded uniform. He told me that I looked 'scruffy', 'unprofessional' and had 'failed to make a good first impression'. I tried to lighten the mood by assuring him that this was just our winter testing apparel, and we'd all be thoroughly logo-ed up in ITV F1 kit by the time of the first race. He muttered under his breath, told me where to find Coulthard, and stalked off.

By necessity we maintained a professional relationship over the following years, but it was another argument that gained me access to the McLaren garage. I was standing in one of the pit-lane walkways between the garages during a race, a useful spot I still use today, from where I could see the McLaren mechanics getting ready to receive one of their cars. James Allen threw down to me and I did a report along the lines of, 'McLaren are preparing themselves, they're getting ready for a pit stop, so this is the time for Ferrari to react.' All fairly standard stuff. After the race Ryan marched up to me. 'What the hell were you doing?' he demanded. 'You gave away information to Ferrari by saying that we were going to pit one of our drivers.' 'Well, I was just standing in the gap there,' I said. 'I could see your mechanics were getting ready. It's not difficult. If I could see it, Ferrari could see it too.' 'Well I'd rather you didn't do that,' he snapped. I didn't back down: 'Why shouldn't I? It's a good line.' Ryan grumbled in that way I had become familiar with, but then said something I didn't expect.

'Alright,' he stuck his hands on his hips, 'I'm going to offer you a deal. I will allow you to stand in the corner of our garage during every race and qualifying session. You will be the only non-McLaren person there. I will even make a little spot for you between the partitions, so that you can be behind the red line, but see everything you need to. But this way you will see stuff in the garage that is privileged information. The deal is, you can report on anything that you would have been able to see from your spot between the garages, but you will not broadcast any information that you have gleaned through being in this privileged position in our garage that might compromise our race operation. And in return you'll have a spot there to talk about anything outside of that.' I knew instantly I'd be able to see more and understand more about the race from inside the McLaren garage, even if I wasn't able to report all of it. The upside for Ryan was revealed in his next words to me, 'If this stops you saying anything about our impending pit stops, that will be better for us.'

I thought about it for a beat before agreeing with a condition of my own that I be allowed to leave via the back of the garage and re-enter if there was a big story at another team that I had to follow up. 'OK, but it's not a bloody revolving door, mate,' said Ryan. I chuckled. The deal was done, and from then on there was a place for me in the far right-hand corner of the McLaren garage. The mechanics even offered me cold drinks at hot races! Thanks to that arrangement I had front-row access to what unfolded in many a key moment in F1 and in particular that explosive Hungarian qualifying session.

Hamilton had been told that when he and Alonso left the pit lane, he should allow Alonso to go in front of him in order that he could burn off some unwanted fuel. At the time, the rules made cars carry the fuel they needed for their first race stint in qualifying, but they could burn off some of this fuel before their qualifying lap because it would get topped up again before the race. However, Kimi Räikkönen came out of the pits at the same time, and in the moment Hamilton reasoned that it would be better for both McLarens that the Ferrari not overtake, so he ignored the instruction. All Fernando knew was that he was slower on that lap because Lewis hadn't done what the team had asked and allowed Fernando to burn off some fuel, meaning his car was heavier and therefore slower. He was furious but managed to set provisional pole on his next run.

With only two minutes and 15 seconds left before the end of the session, Alonso was fastest. He came in first, his tyres were changed, and then he was given a visual countdown by the chief mechanic and lollipop man Pete Vale to re-enter a clear position on track. Meanwhile, Hamilton arrived behind Alonso and had to wait for his tyre change. Alonso was held for about 12 seconds, before Vale lifted the lollipop and told him to go, but Fernando sat there, not going anywhere. By the time he finally moved off and Hamilton could drive forward into the McLaren pit box for his new tyres there was only one minute and 25 seconds of the session remaining. An 'out' lap took one minute 30 seconds. It had been made impossible for Hamilton to get round in time to start his final lap, and thus he was unable to improve on his qualifying time. The session ended with Fernando on pole, and Lewis in second place.

Had Fernando deliberately screwed over Hamilton by making sure the time ran out and he couldn't get his lap in? Ron Dennis certainly thought so. The McLaren boss threw his headphones across the pit wall desk and marched straight to Alonso's physiotherapist Fabrizio Borra, who doubled up as the pit board man for Alonso when the cars were running on track. Borra was clearly shocked by Dennis's intervention, putting his hands up in a 'don't shoot me' gesture only for Ron to march him firmly down to the parc fermé area at the pit entrance, where Dennis knew Alonso and Hamilton would be shortly arriving.

Egged on by James Allen, avidly following events in the commentary box and, as my pit lane predecessor, frustrated not to be able to get in there and ask the questions himself, I dived in for a live interview with Ron, matching his pace as he strode along and ignoring the fact that he clearly didn't want to talk to me: 'What happened there?' Ron was visibly seething, and with his hand still firmly on Borra's shoulder said flatly, without slackening his pace or meeting my eye, 'We'll discuss it later within the team, and we'll sort it from there.' I followed up with a second question, but he wasn't budging. As a former racer, co-commentator Damon Hill was partly on Alonso's side: 'That's sport. That was tactics and timed to perfection,' he said over commentary. 'Alonso really, I'm afraid, stuffed it to Lewis.' 'I feel very strongly,' he added, 'that racing drivers should be allowed to race. Alonso did a little bit of a naughty thing there, it has to be said, but it was good tactics.' It was an observation borne of experience; Hill having been on the receiving end of a few 'naughty things' during his own time driving in F1.

Louise Goodman was covering Hamilton post-qualifying, so I headed to interview Fernando. The explanation he gave me that afternoon was that he could see Pete Vale telling him to go, but he was actually listening to a different count from his race engineer Mark Slade. 'We wait for the countdown in the radio, and we go,' said Fernando. 'Sometimes it's 10 seconds, sometimes 45 seconds, like the first stop of today. Sometimes 10 or 15, like the second. But I think the calculation was wrong, because my teammate didn't complete the lap and I crossed the line by two or three seconds. So it was really tight. These things unfortunately happened today to us.'

When I asked why he hadn't responded to the lollipop going up, Fernando had a ready explanation: 'Sometimes the mechanics, when they see another car [waiting] behind, they try to change the tyres and go. But I have the radio, 5-4-3-2-1, so I wait for the zero.' OK, I thought, but his team boss wasn't buying that. 'It's unfortunate that some people will jump to the conclusion that you did it deliberately,' I countered, to see how he'd react. 'I know,' replied Fernando, 'because I think the TV shows Ron not very happy, but I think he's not happy with the engineers, as they do the

When Capital Radio issued me with a press pass in July 1996 I had no idea that it would lead me to a press conference with Damon Hill, a meeting with James Allen and my first job in Formula 1.

PRESS CAPITAL 95·8 FM

TED KRAVITZ
REPORTER
EXPIRES END JUL 98
IRN 002840

Murray's last race commentary for ITV, USA, 2001. When Murray and Martin were in full flow they used to sway together like windscreen wipers. Murray remained keenly in touch with the sport until his death in 2021, aged 97.

The ITV F1 team on Murray's last weekend. From top row to bottom, left to right: Jenny Bozson, Gerard Lane, Geoff Kay, Dave Boyd-Moss; Mat Bryant, Dave Hill, Kevin Chapman, Rob Walker; Les Horne, Bill Fievez, Tracy Rooney, Ron Trickett; Jon Pearce, Di Finch, Sally Blower, John Tidy; Kevin Piper, me, Andy Parr, Jo Hybert, Steve Aldous; Mark Blundell, Tony Jardine, Louise Goodman; James Allen, Jim Rosenthal, Murray Walker, Martin Brundle.

Michael Schumacher demanded excellence from everyone around him. Interesting, well-considered questions were rewarded with an insightful answer. Jonathan Legard and Maurice Hamilton listen in behind, as Michael's media manager Sabine Kehm keeps an eye on proceedings.

The six-car race, Indianapolis 2005. The day we ripped up our running order and watched helplessly as Formula 1 dealt its popularity in the USA a heavy blow.

All smiles between McLaren's Fernando Alonso and Lewis Hamilton before an explosive 2007 season characterized by threats, professional fouls and industrial espionage.

Putting the tumult of 2007 behind him, Lewis Hamilton, surrounded by his father, brother and the McLaren team, celebrates the pure joy of clinching the 2008 world championship on the last corner of the last lap of the last race.

Jenson Button won the 2009 world championship in a Brawn car that was run on a shoestring but had the genius double diffuser from the start.

One of my frequent BBC interviews with Lewis Hamilton, but there wasn't much to smile about from 2009 to 2011: three seasons where Jenson Button and Sebastian Vettel shared the championship spoils.

Carlin mechanics attempting to cheer up Sebastian Vettel at Spa-Francorchamps, 2006, the day after he sliced the top off his finger. Vettel subsequently showed that famous finger on the podium every time he won.

Right Socially distancing from Valtteri Bottas. Formula 1 was the one of the first global sports to resume during the summer of 2020. We operated effectively in a bubble at the circuits, with no fans present, but the TV coverage brought a bit of normality and excitement to viewers kept indoors during Covid-19.

Right One of the Covid precautions distanced reporters from the pit lane. My trusty monocular was the only way to see what was happening in the garages.

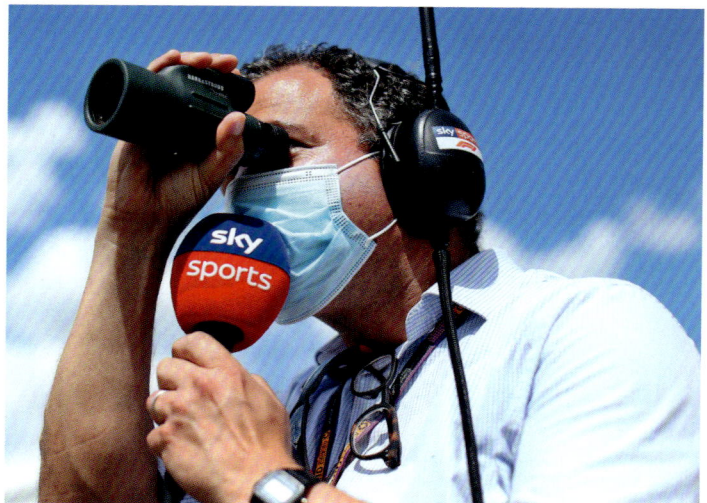

An interview with W Series runner-up Alice Powell, 2019. All the W Series drivers were seriously quick and were pathfinders for the F1 Academy, which now seeks to discover the next female F1 driver.

Lewis Hamilton and Max Verstappen's intense rivalry ended with a race that damaged faith in Formula 1's referees.

Hamilton was defenceless against Verstappen's new soft tyres. Had the race been run to the rules as they were understood, the 2021 Abu Dhabi GP would have finished under the safety car and Lewis Hamilton would have been an eight-time world champion.

Azerbaijan, 2024. Even when things aren't going his way, Lando Norris has an easy charm that makes him a fan favourite.

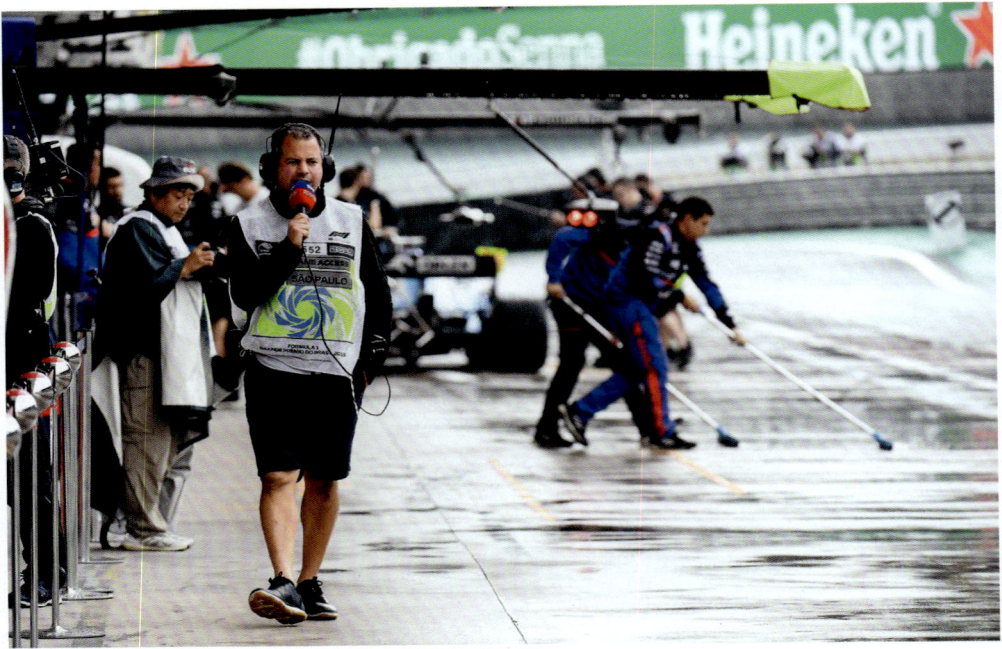

Insomuch as there is a uniform for the pit lane reporter, I favour navy-blue shorts (which don't get as wet as trousers in the rain), and in hot weather, when permitted, my trusty Birkenstocks.

Flying wingman to George Russell in an RAF Typhoon.

calculation'. What Fernando didn't explain, but which later emerged, was that he had been having an argument on the radio with his engineer about why he'd been given used, hard tyres for his last run, rather than the quicker new soft tyres.

This was all just about plausible, but the matter was far from over. Alonso, Hamilton, Dennis and Ryan were called to talk to the FIA stewards on Saturday afternoon to explain themselves, in an unusual investigation for an intra-team dispute. Some hours later the stewards showed that they weren't buying Alonso's explanation either, and found him guilty of an act 'prejudicial to the interests of the sport or competition.' He was demoted five places to sixth on the grid, and Hamilton inherited pole.

Like the stewards, most people inside F1 felt sure that it had been a bit of gamesmanship from Fernando. His former Renault engineer Pat Symonds even used that word when I spoke to him on Sunday. On our pre-race show we analysed the stewards' judgement that it had been a deliberate ploy by Alonso. In what had become his regular post-qualifying interview with Steve Rider, Hamilton said that he didn't think Fernando had a great excuse for what he did. Ex-driver and ITV pundit Mark Blundell echoed another popular view in the paddock – even if Fernando had stitched Lewis up deliberately, it was a team-on-team incident. No other team was disadvantaged, so why did the stewards get involved at all?

The feature I made covering the story on race day opened with 'McLaren say this is more a case of cock-up than stitch-up.' Looked at that way (and the stewards had acknowledged the role McLaren had played in the mess, docking their constructors' championship points for the weekend), I'm willing to believe Alonso's explanation that he was waiting while he argued about why he had been given the scrubbed hard tyres, and the countdown, because in the camera shot you can see him looking over to Slade, his engineer, on the pit wall. There was an unlikely theory at the time that he might have been looking over at Fabrizio Borra, waiting for some kind of hand gesture signalling the perfect time to go, allowing him time to start his own lap but denying Hamilton the opportunity to do his. This falls down over the fact that there's no way that Alonso would have been able to see Borra, and even if he had, Borra wouldn't have been able to count the exact seconds. And, anyway, how would the two have known to pre-plan that before the session?

To this day, Alonso insists that his primary motivation was to follow the countdown and discuss tyres with his engineer over the radio, and not to stitch up Lewis. That there could have been so much doubt as to what really happened must seem surprising to a contemporary audience used to hearing full, unedited team radio exchanges between driver and engineer, but this was at a time when team radio was a closely guarded secret, digitally encrypted and not open to TV broadcasters (unlike today, where the regulations stipulate it must be accessible). I think Fernando didn't care that Lewis was behind him. I think he would have been perfectly happy to have a discussion with Slade about his tyres even at such an unusual and untimely moment, notwithstanding the fact that it might mean Hamilton struggled to get his lap in. If Alonso had done it deliberately, the mental agility required to have counted to the second precisely how long there was in the session and judge how long he needed to remain stationary to run Hamilton out of time, but not himself, would have been extraordinary. But it is possible that Alonso was thinking, 'Well, he screwed me over earlier on, so I'm quite at liberty to screw him over now.' And that would explain why Lewis said, 'I don't think Fernando has a great excuse for what happened today.'

The broader context for all this, of course, was Fernando's increasing frustration that his role as F1's kingpin was being challenged, and his sense that McLaren's management weren't doing enough to help him in his quest for a third world championship. Meanwhile, there was another story afoot, a story that turned out to be one of the most shocking and dramatic episodes to occur during my time in Formula 1.

In 2007, the Ferrari dream team was changing. Michael Schumacher had retired, Ross Brawn had followed him out of the door, and there was another key player who felt things weren't how they used to be. Nigel Stepney was a British mechanic-turned-team-manager who was known within Ferrari as 'The Enforcer'. He had been brought in by Brawn to help get the Maranello race team into shape, and he performed his role with military-grade efficiency. He was an imposing figure. Some people found him a bit scary, and he certainly had a ruthless streak, but he was extremely effective. He had learned Italian fluently, which helped him gain the respect of the mechanics. Although he was admired by many within Ferrari, he had also been passed over for promotion and felt he wasn't being properly appreciated.

Stepney had a friend at McLaren, chief designer Mike Coughlan. Coughlan had landed a plum job at the team's base in Woking largely off the back of a particularly good car he designed for the Arrows team, the A23. Despite his past success Coughlan found it difficult to flourish at his new team, so he, together with his friend Stepney, decided to offer their joint services as a kind of 'dream team' to the Honda F1 team, who were recruiting. It was around that time that something extraordinary happened. Stepney gave Coughlan a 780-page dossier detailing various parts of Ferrari's soon-to-be championship-winning car. Design files, drawings – a copy of what was in effect the 'owner's manual' for the Ferrari F2007. It was a detailed guide to the car and how to run it.

Coughlan said that he took it out of 'engineering curiosity', but somewhat naïvely then asked his wife Trudy to take the documents to his local print shop near the McLaren factory in Surrey to be digitized and copied on to compact discs. This was not a small job – it was almost 800 pages of confidential Ferrari information filling two binders. What the Coughlans couldn't have foreseen was that the manager of the print shop was not only an F1 fan, but a Ferrari fan. Noticing the prancing horse logos on car blueprints, the manager grew suspicious and began to realize the significance of what he'd been given to copy on to disc. Among the information was an email address. Following up on his suspicions that something wasn't quite right, the copy-shop manager sent an email to the address with his concerns. It was none other than Ferrari's then sporting director Stefano Domenicali. I like to imagine that Domenicali's reaction on reading the email was to spit out his morning cappuccino; he certainly took the information seriously – immediately contacting Ferrari's internal security team and the police, who took the matter from there. As this intellectual-property scandal duly erupted, it quickly became known as 'Spy-gate'.

McLaren was investigated by the FIA in an effort to determine to what extent the information in Coughlan's trove had been disseminated within the team. McLaren's response was that only a 'rogue' employee had possession of the documents, that the information had not been shared around the factory, and that it hadn't been used to any sporting advantage. This first FIA investigation therefore cleared McLaren of significant wrongdoing because while its chief designer had possession of the stolen information, the FIA could not prove that anything had been used to the team's benefit. Ferrari team principal Jean Todt was furious with the decision and promised to seek a judicial review.

This was all happening at the time of Hamilton and Alonso's pit lane bust-up. According to Ron Dennis, on the Sunday morning after the qualifying incident, Fernando Alonso came into his office in the McLaren motorhome. Alonso was furious that, having been recruited as world champion, he was not being given the advantageous treatment over Hamilton that he believed was his due. Dennis, it's worth noting, has always denied that he had promised Alonso number-one driver status.

As tempers rose, Alonso told Dennis of email documents he possessed that could be damaging to McLaren in their continued 'Spy-gate' investigation. This was the first Dennis knew of emails between Alonso and others at McLaren, which proved that there were more people within McLaren who had been aware of the Ferrari information than had been admitted to the FIA. Dennis called his deputy Martin Whitmarsh into the meeting where, it was alleged, Alonso threatened the McLaren duo that he would go to the FIA with these emails if they didn't do what he wanted. Rather than be dictated to by his driver, in a pre-emptive move Dennis called FIA president Max Mosley to inform him about the new disclosures. Later Alonso's manager told Ron that Fernando was sorry, that he had changed his mind and retracted what he had said. But by then it was too late – the call to Mosley had been made.

In any case, it turned out Max Mosley already knew about the emails – Bernie Ecclestone had told him about them (it was presumed Bernie's information had come from Flavio Briatore, Alonso's manager and confidant).

A much more detailed investigation than the first was then launched by the FIA. It confirmed the existence of the emails between Coughlan and McLaren test driver Pedro de la Rosa, and then between De la Rosa and Alonso, talking about such things as what gas Ferrari was using to inflate its tyres, and the F2007's weight distribution and ride heights – the sort of information that could be very useful for a rival team like McLaren.

As we covered the unfolding story on ITV F1, I became the 'Spy-gate' correspondent, updating the saga with features each race weekend. It was an exhilarating period as more information emerged, each snippet more incredible than the last. I suppose, like the frog in the pot of water, I didn't realize at the time what a big story it was becoming, because from my point of view it was new revelations and surprises, week after week. What frustrated me was that a lot of people in the paddock quickly tired of the story, and started dismissing it in terms of 'Oh, it's just F1 politics.' But it wasn't just politics, it was about unfair advantages. Ferrari had a highly justified grievance about the fact that its intellectual property had not only ended up in McLaren's hands but might also have been used to make the McLaren quicker.

The result of the FIA's second investigation and the verdict of the World Motor Sport Council came during the Belgian GP weekend in September. The McLaren team was found to be guilty of using and benefitting from Ferrari's IP and was punished by being excluded from the 2007 constructors' championship, and handed a staggering $100 million fine (although half was suspended). Mosley's view at the time was that looking back on the story 20 years hence, history would probably show that the FIA had been, if anything, too lenient on McLaren, and that not only should they have thrown Hamilton and Alonso out of the 2007 drivers' world championship, but they should have thrown the team out of the 2008 championship as well, because the information gleaned could have been used on the following year's car. McLaren would then have been able to start 2009 cleanly.

What complicated the picture was the well-known, long-running personality clash between Max Mosley and Ron Dennis, something Max talked about when the subject of the $100 million fine came up in Michael Shevloff's 2020 documentary *Mosley: It's Complicated*. Dennis always felt the aristocratic Mosley discriminated against him because of his humble family background and the fact he started in F1 as a mechanic. Mosley said that what really annoyed him about Spy-gate was that Dennis hadn't been straight with him on who in McLaren knew what and when.

In the fallout after Spy-gate McLaren had its shot at the 2007 constructors' title taken away, a championship it would have won. The team then contrived to lose the drivers' title as well. At the penultimate race, in Shanghai, Lewis Hamilton skated into a gravel trap at the pit entry having been kept out too long on old, wet tyres, which allowed Räikkönen to win the Chinese GP. This put the Finn up against the McLaren drivers in a three-way fight at the final race of the season in Brazil. Against the odds and with the help of Ferrari teammate Felipe Massa, Räikkönen got the job done, and clinched the world championship with 110 points to the

109 of Hamilton and Alonso. In the end Fernando had been right. The principle of supporting both drivers ultimately lost McLaren the drivers' title.

What was certainly clear after Spy-gate was that the Alonso-McLaren relationship was over. His contract was cancelled, and after just a year with the team he returned to his former home at Renault, leaving Hamilton clear to mount another championship challenge at McLaren in 2008. What a season. And little did we know, but the following year would see yet more drama surrounding the two men.

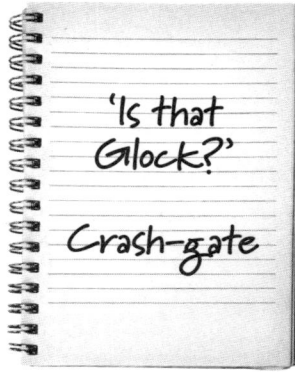

'Is that Glock?'

Crash-gate

Chapter 12

A Fifth and a Fix

———

Sports commentators are a talented bunch. What makes a great commentator? I don't think there's a formula, but one important skill is the ability to come out with memorable lines, delivered completely off the cuff in the moment, and if you're lucky, come to be inextricably linked to the sport in question.

'Some people are on the pitch. They think it's all over. It is now!' was a line of genius from Kenneth Wolstenholme in the closing moments of the 1966 World Cup Final. In F1, Murray Walker had a gift for the right line at the right time. 'And colossally, that's Mansell,' when Nigel had his puncture in Adelaide in 1986, or 'I've got to stop because I've got a lump in my throat,' voicing the feelings of a nation when Damon Hill won the world championship in Japan a decade later. More recently, David Croft's 'Through goes Hamilton!' at the 2022 British GP was a reactive line judged and delivered to perfection.

However, there was one snippet of race commentary delivered by Martin Brundle on the last lap of the 2008 season that was devastatingly simple, and yet became iconic, forming part and parcel of F1 lore: 'Is that Glock?' That simple question (and the confirmation that it was, indeed, Timo Glock, going slowly) encapsulated the dramatic end to Lewis Hamilton's first successful world championship campaign.

That 2008 season was a busy one. On top of an 18-race F1 calendar, Louise Goodman and I also presented ITV's coverage of the British Touring Car Championship, which meant we spent every weekend at a racing circuit. We'd go from Australia to Oulton Park, Snetterton to Singapore, and Budapest to Brands Hatch. It was great fun. I'd been a fan of the BTCC since Murray started voicing the highlights on BBC's *Grandstand* and had fond memories of Andy Rouse battling Frank Sytner, Nigel Mansell's escapades in a Ford Mondeo and Rickard Rydell's incongruous Volvo estate. Our Sunday coverage on ITV4 was, to say the least, comprehensive. It grew to encompass seven hours of live television from the opening race of the day for the Ginetta Junior Championship to the third and final BTCC round of the meeting. Incidentally, if you're hoping to spot the next F1 star, Ginetta Juniors is a great category to keep an eye on. With a minimum age of just 14, it is the first car series that many drivers graduate to on their way from karting up the racing ladder. Years later in 2014, when Steve Rider had taken back presenting duties on ITV's BTCC show, he passed on a tip – the only one he ever gave me – about a promising driver: 'Watch out for this kid in Ginetta Juniors called Lando Norris. He looks something special.' I'd worked with Steve for years and had never known him to single someone out like that before. Or since, for that matter, but Steve knows how to pick them.

In 2008, reeling from being thrown out of the previous year's constructors' world championship, McLaren was out for redemption. Having technically been classified last, the team was obliged to run the two car numbers of 22 and 23 at the bottom of the entry list (this was before drivers were allowed to choose their own race numbers). The team was at least saved the indignity of being relegated to a garage at the end of the pit lane after a moment of kindness (or pragmatism) from Bernie Ecclestone, who acknowledged that it would probably be better for the show and the TV coverage if the top teams competing for the world championship were near each other in the pit lane rather than separated by 10 other garages. The story went that Bernie had planned to give McLaren the third garage, but Renault objected. So they ended up in the fifth, next to Red Bull Racing. With Fernando Alonso gone, Heikki Kovalainen was brought in as a low-friction teammate for Hamilton, leading to a much more harmonious atmosphere in the McLaren camp.

The 2008 McLaren MP4-23 was a straightforward development of the car that should have won the previous year's constructors' title, and Lewis duly kicked off the season in style, winning from pole in Melbourne. Over at Ferrari a strange handling characteristic was taking the edge off Kimi Räikkönen's title defence. James Allison, one of Räikkönen's ex-technical directors, once described the Finn as having 'soft hands'. Not literally, unless Allison had some insight into Räikkönen's manicure routine that I'm unaware of, but in his particular sensitivity to a car's balance.

Too often Räikkönen found that he couldn't get his front tyres to work, losing him time as the car understeered into corners, the front end never biting and gripping up. A mid-season upgrade actually developed the Ferrari F2008 even further away from Kimi's style and towards the preferences of teammate Felipe Massa, so it was the Brazilian who emerged as Hamilton's main title rival. Felipe was outstanding that year. He was driving less aggressively than in previous seasons and made fewer mistakes. Had the car been more reliable, he could have won the 2008 world championship comfortably. However, retirements from the first two races, an engine failure in Hungary and a botched pit stop in Singapore (more on that later) meant that Hamilton edged ahead, and Massa was left on the back foot going into the title decider in Brazil. Hamilton had a seven-point lead. In the days of the old 10-8-6-5-4-3-2-1 scoring system it was easy to work out that fifth place would guarantee Lewis the title even if Massa won the final Grand Prix.

Massa had a big home advantage at the decisive race of the season. His family and friends were in the paddock, and the crowd chanted his name throughout the weekend. There had even been support from the presenters of a Brazilian TV show that played practical jokes on people and filmed their reactions, a fashionable genre at the time. These two pranksters gatecrashed a McLaren sponsor function before the weekend began and handed a fluffy black toy cat to Lewis – black cats being a symbol of misfortune in Brazil – hoping to put some kind of curse on his race weekend. Hamilton laughed it off, pointing out that in the UK having a black cat cross your path is a sign of good luck.

No cats appeared, fluffy or otherwise, in ITV's final pre-race build-up. Louise and I flipped a coin to decide who was going to do which pre-race feature on the two contenders. I landed Massa. My piece started with the thought that Brazilian racing drivers had a lot to live up to (cue shots of Ayrton Senna), but that Massa had harnessed the adulation to raise his game. 'I've taken all the good energy around and exchanged it for extra power in the car,' he said, before we heard from former Brazilian champion Emerson Fittipaldi about how Felipe had matured. I ended the piece with an idea that I'd had the week before, although I hadn't quite figured out how it was going to work.

Knowing that the McLaren team had been carefully keeping a low profile in Brazil, travelling to and from the track in plain clothes (not unusual in São Paulo), I had the somewhat quixotic idea that I should see what happened if I deliberately attracted some attention. On Thursday I asked Massa what he thought would happen if I went out among the hugely partisan Interlagos crowd as a British TV presenter displaying a Union Flag. 'For me, it's not a good idea,' he cautioned. Despite this incredibly sound advice it was ITV's last F1 weekend before the BBC took over, and there was an end-of-term feeling in the air, so I decided to go for it. I stood on a wall overlooking the grid and did a little piece to camera next to the fans, separated from the crowd by a chain-link-fence which in retrospect probably saved my bacon given what I did next. I unfurled a modest Union Flag that I'd borrowed and waved it about, to see what would happen. For a second, nothing – astonishment, no doubt – but then the boos and whistles rang out, and someone in the third row lobbed a plastic cup half-full of (very tasty Brazilian Antarctica brand) beer over the railing and on to my shoulder. It was a great shot by the spectator, which is more than you can say for the TV piece.

On this occasion my attempt to keep things entertaining with the crowd's help may have gone slightly awry, but it's also true to say that

I've had various hit-and-miss interactions with crowds of fans across my Formula 1 career. After the 2022 Australian GP the crowd ended up using me as a prop. Joining the post-race track invasion to give viewers a sense of the energy and enthusiasm of the Aussie crowd for Charles Leclerc's win, I was handed a shoe, by a fan, filled with warm beer, and encouraged to drink. As the crowd clearly sensed, while on live TV, I didn't really have the option to back out (and thanks to Daniel Ricciardo I did have some awareness of this Australian sporting tradition of the 'shoey'), with the result that viewers at home were treated to the unedifying spectacle of me downing the shoe's contents. And do you know what? It wasn't that bad.

Anyway, back to Brazil, and with my impertinence duly punished, I was fortunate that the beer was rinsed away by a heavy rain shower just before the start. This led to a delay so that the mechanics could fit intermediate tyres to the cars. When the race started, Massa led from the off, and as conditions dried out he kept the lead through the switch back to slicks. The fact that Hamilton didn't need to win outright – only finish fifth – understandably led to a cautious approach by the McLaren engineers. As I watched from the garage the team seemed to build a little inertia into their strategy calls, taking extra time to think through every decision. Lewis was a couple of laps late changing on to slicks, but he regained that vital fifth place on track. And there he stayed until, with 10 laps left, the rain came again, lightly at first. As everyone dived into the pit lane for a quick tyre change the Toyotas of Timo Glock and Jarno Trulli stayed out, the team having judged it wasn't damp enough for wets. That elevated Glock to fourth.

Like the rest of the field Hamilton pitted for intermediates, with frontrunner Massa stopping a lap later. The Ferrari had a clear track ahead, and Felipe looked much more comfortable in the rain. Hamilton rejoined the track in fifth, now behind Glock, who had yet to come in. Coming up fast in sixth was Sebastian Vettel, who had proved, by brilliantly winning the Italian GP four races earlier, that he and his Toro Rosso made for a very quick package in wet conditions. Suddenly the threat to Hamilton was Vettel closing in behind. At a moment when Lewis went slightly wide, the German youngster overtook him. There were gasps and then shouts of frustration in the McLaren garage as the timing screen updated itself: Hamilton was now in sixth.

The layout of the McLaren garage was the same at every race. Mechanics sat in their fire suits and chrome stormtrooper-inspired helmets in an arc around a TV on each side wall, with tyres stacked in the middle where the drivers' families were allowed to stand and watch. (There was a viewing gallery in the middle of the back wall, but that was full of sponsors and guests.) From my spot in the front right-hand corner I couldn't see a TV screen, but I was listening closely to James Allen and Martin Brundle. Over my left shoulder I could see Lewis's step-mum Linda with his girlfriend Nicole Scherzinger alongside. They were bouncing on tip toes, hands either clenched by their sides or covering their eyes. Lewis's brother Nicholas sat with the mechanics, eyes riveted to the screens. Lewis's dad Anthony was not visible, as he preferred to watch from the computer racks behind a partition wall.

There was a lot of praying going on. Down at Ferrari, Felipe Massa's mother Ana, father Luis and brother Eduardo were cautiously optimistic. Felipe had kept the lead, and it wasn't raining hard enough for Glock's dry-weather tyres to lose grip just yet. I felt a chill run through the McLaren garage as everyone there faced up to the very real possibility that they were about to lose the world championship for a second year running. The 71st and last lap came, and I suddenly realized another reason I was cold. The wind had picked up in that way it does just before a rain shower – and at Turn 4, the Lake Descent, it was raining again, hard. As Massa came out of the last corner to win the Brazilian Grand Prix, Hamilton had just entered the twisty section of the track. Massa had done what he needed to do, and his family erupted with joy, only to be shushed by clued-up Ferrari mechanics, who knew the championship wasn't over until Lewis crossed the line. I looked across the pit lane, watching Ron Dennis on the McLaren pit wall, frozen to his seat, eyes glued to the TV screen, as I listened to Allen and Brundle in my headphones.

> James: 'Can Hamilton do anything? Can he run it up the inside of Vettel? Only a few corners to go now and desperation starts to creep in for Lewis Hamilton.'
>
> Martin: 'Räikkönen's third and . . . Is that Glock, is that Glock going slowly?'
>
> James: 'It is!'
>
> Martin: 'That's Glock!'

It was true, Glock's dry-weather tyres were suddenly useless in the wet conditions. He had no grip. Hamilton scythed past him.

> James: 'Oh my goodness me! Hamilton's back in position again. A hundred thousand local hearts sink in the grandstand. It's handed the place back to Hamilton. He comes through. And

I'm sure that he's going to claim fifth place, which is all he needs to do to become . . .'

Martin: 'Yes!'

James: 'The 2008 F1 world champion – Lewis Hamilton! Well, the Ferrari boys are celebrating, they think they have won.'

Martin: 'They're wrong. You're wrong, guys.'

James: 'They absolutely haven't won. Hamilton finished fifth. . . . You will never see a more dramatic conclusion to any motor race, let alone a Grand Prix, than that. And the result of it all is that, in the most harum-scarum way possible, he doesn't make it easy for himself, does he? Lewis Hamilton is the world champion.'

Martin: 'Unbelievable.'

Hearts and heads sank in the Ferrari garage. The last corner of the last lap of the last race of the whole 2008 world championship! One Ferrari mechanic head-butted the garage wall so hard that it shattered. Hamilton himself didn't know if he'd done enough. He came on the radio: 'Do I have it? Do I have it?'

For Lewis to wait for confirmation of whether the championship was his must have felt like an eternity, but in fact took five seconds or so for his engineer Phil Prew to check with sporting director Dave Ryan, who gave the thumbs-up for Prew to confirm it on the radio. Hamilton had won the world championship by a single, solitary point. The McLaren mechanics already knew he had it, and they were on the pit wall cheering him home while the garage emptied into the pit lane, everyone screaming their heads off in pure joy. By that time cameraman Andy Parr had joined me, and we looked at each other, laughed and shook our heads in disbelief as we walked across towards Ron Dennis and Martin Whitmarsh on the pit wall. It was Hamilton's day, but Massa won just as many plaudits for the impeccable drive that had won him the race and the dignified manner in which he conducted himself once he had been told over the radio that the title had slipped out of his grasp.

As I mentioned, Felipe Massa lost valuable points at several key races before that day in Brazil, infamously at the Singapore GP, which later became known as the 'Crash-gate' race, but we would have to wait almost a year before the true facts came to light. Reporting on the race-fixing scandal for the BBC I called it 'one of the worst cases of cheating in sporting history'. I had debated with myself whether to word it so strongly, but the facts were undeniable, despite what we later found out about the timeline of events and the circumstances surrounding them that provided some mitigating context as to why those involved acted the way they did.

After his ill-fated 2007 season with McLaren, Fernando Alonso had returned to the Renault team where management were keen to show him that they were heading back to race-winning form in the hope they could keep their mercurial champion long-term. The Renault was a decent enough car, but it wasn't a winner, much to the disappointment of Carlos Ghosn, the Renault Group president and CEO.

Ghosn had a reputation for not being the most committed CEO among the F1 manufacturers and certainly (at least when it came to budgets) the most frugal, and re-iterated his demands that Renault perform 'at the top level' of F1 or the company's investment would have to be reviewed. That pressure to win was passed down from Ghosn to Briatore and on to the race team, which was headed by executive director of engineering Pat Symonds.

With Hamilton and Massa still fighting for the world
championship, the F1 circus left Europe and headed for
Singapore and the sport's first ever night race. In free practice,
there was a surprise for everyone in the pit lane: the Renault was
actually quick. Alonso immediately clicked with the technically
and physically demanding track, and once the engineers had
made some adjustments, he was fastest across the race
preparation runs of FP2, and the qualifying practice runs of
FP3. The car was kind on its tyres and had good mechanical grip.
Out of nowhere, Alonso was suddenly a real contender for a
race win. But come qualifying, he had an unfortunate setback.
A fuel-flow problem to the engine meant that his Renault came
spluttering to a halt on track, and he couldn't set a time in Q2.
Fernando was left watching helplessly as slower cars qualified
ahead of him, and he ultimately wound up 15th on the grid.
Alonso, Briatore, Symonds, chief engineer Alan Permane and the
rest of the team were absolutely gutted. The car was worthy of
running at the front, but was now out of position, starting three
quarters of the way down the grid. It was an injustice – but given
the fault was a technical glitch the Renault personnel had no one
to blame but themselves.

Starting just behind Alonso on the Singapore grid was
his teammate, Nelson Piquet Jr. The Brazilian didn't have
the pace of Alonso and had failed to get out of Q1. It was an all-
too-familiar story for the likeable Piquet, whom I'd got to
know a little during his 2006 GP2 season as a result of
all those interviews about his championship battle with
Hamilton. Piquet was a quick driver. He had won the British F3
Championship and finished second to Hamilton in GP2,

but then spent 2007 on the sidelines as Renault's test and reserve driver. He had entrusted Flavio Briatore with managing his career, which paid dividends when he got the Renault race drive for 2008. However, he had a poor start to the season, retiring from six of the first nine races. But in Germany, he finished second, thanks to a lucky bit of pit-stop timing and the intervention of the safety car.

In 2008, there had been a significant change to the rules. The FIA weren't happy with the way that, as soon as someone crashed and it looked like there would be a safety-car period to slow the cars down, everyone would race hell-for-leather back to the pits to get their pit stop done without losing too much time. So race director Charlie Whiting inserted a rule that as soon as the safety car was deployed, the pit lane would 'close' (it wouldn't physically close, it just meant that if a car did come into the pits while the safety car was deployed it would get a time penalty). Once everyone was proceeding at a safe slow speed behind the safety car the pit lane would then 'open', allowing everyone to dive in. However, trying to solve one problem had created another with new dangers in the pit lane, one example of which would be crucial in deciding that year's world championship.

What Piquet and Renault learned from the German GP was that, should you have a lousy qualifying and start well down the grid, if you then happened to make a pit stop and fuel your car to the end of the race – and if someone then happened to crash and the safety car be deployed – you might well suddenly find yourself at the front, as the cars ahead of you make their pit stops. From that point, it would be a case of defending your newly found place at the head of the field, which is how Piquet went from 17th on the grid to second and the podium in Germany. It was with this experience in their data memory banks that Renault sat down after qualifying in Singapore and thought hard about how to get Alonso to the front of the field, where he would be able to run at his true pace.

According to the evidence that emerged in the subsequent investigation, this is how things played out that Saturday night and Sunday. With Hockenheim in mind, Pat Symonds reasoned that if he could guarantee the timing of a safety-car period, Alonso could pit just before it, fuel up, gain track position at the front and use the inherent pace in the car to stay ahead and win. But how can you guarantee a safety-car period? Symonds figured that a crash in an area of the track that was hard for the marshals to reach would do the trick. Perhaps against his better judgement, Symonds came up with a plan to use Renault's second car to cause a safety-car period. He asked Piquet to crash. Under obvious pressure to keep his drive and please the team, the Brazilian agreed. The exit of Turn 17 was selected as the perfect spot, as it was difficult to access and recover a car quickly, a safety car being the certain consequence.

Whether in a coincidental foreshadowing of what was to come or because he was practising, Piquet spun his car on the formation lap to the grid. Once the race got underway Alonso made that early pit stop on lap 12, a move that made no sense at the time as generally those starting out of position, as he was, aim to run long in an attempt to work their way up the order. Then, on lap 14, coming out of Turn 17, Piquet spun and crashed, perhaps more heavily than planned, into the inside wall. The safety car was duly deployed, and with others pitting, Alonso eventually took the lead, and kept it. In the moment we thought that the timing of Alonso's stop had been lucky. We had no reason to suspect that there was anything untoward going on. Indeed Fernando went on to win the next race at Fuji, a chaotic rain-hit affair, proving his car was competitive.

And we were none the wiser until the middle of the 2009 season, at which point Piquet was fired by Briatore to make way for Romain Grosjean. Soon after, the Brazilian took the opportunity to come clean, telling the FIA in a sworn statement what exactly happened in Singapore, and an investigation immediately ensued. Coming just two years after the McLaren-Ferrari Spy-gate saga, the affair soon picked up the moniker Crash-gate.

Fixing a Grand Prix result is an incredibly risky thing to do, and the number of people who would need to be in on it tends to mean that eventually the true story comes out. In this case, it emerged that Nelson Piquet Sr had actually told FIA race director Charlie Whiting what had happened in Singapore at the end of the 2008 season in Brazil. Whiting had in turn told FIA president Max Mosley, but at that point the FIA didn't have enough evidence to do anything about it. It was only after Piquet Jr's statement that Mosley employed some ex-police investigators to interview Alonso and Pat Symonds and the truth started to emerge.

Alonso flatly denied knowing anything about the plan. I believed him, and so did the FIA's investigators. There was certainly no reason for him to have been aware in order for the plan to work. He might well have thought that lap 12 was a bizarrely early time to stop and made little sense as a race strategy, but his natural reaction in the cool-down room after the Grand Prix, where he immediately credited the safety car for his win, could not have come from someone who had been in on the plan to fix the result. When I asked him a year later whether he still counted Singapore '08 as a proper

race victory, he was completely unfazed by the question and had no hesitation in answering.

Fernando: 'Yes, I do.'
Ted: 'Even though the team effectively manufactured it?'
Fernando: 'Well, I think that's an interpretation. There are many interpretations [of] how you can win the race. That was in the very early stage of the race, and it was a long race to do. The car was performing well, I did no mistakes, and I still count it.'

Walking away from that interview I was impressed by Alonso's nonchalance. But looking back on it, Singapore is always such a hard race to win that I was willing to accept his point, or rather his 'interpretation'.

The extent of Briatore's involvement was very much open to interpretation, most of it created by the man himself. Piquet alleged that his former team boss was an active participant, and alongside Symonds had asked or instructed him to crash in Singapore. Briatore denied it and came out fighting. I was present at a tense media briefing Flavio gave in the Renault motorhome at Monza in 2009, the race before the FIA hearing. We were ushered in by his long-time media manager and right-hand woman Patrizia Spinelli, and by one of the many Renault company PR men who travelled the world with the team. It was a complete scrum, everyone crowded around a single table, and my BBC cameraman only just got a position.

In came Briatore, resplendent grey hair tumbling down to the collar of his trademark open-necked shirt. He looked like he hadn't slept. He wasted no time in telling us that Piquet was just bitter and angry about being sacked, and that both his own and Renault's lawyers had commenced legal action against both Nelson Piquet Jr and Nelson Piquet Sr in the high court of Paris – for blackmail. The way he left the word 'blackmail' until last, and the portentous way he said it was pure theatre. 'How harmful have the accusations been to Renault's reputation?' I asked ('It's major damage') and then, slightly provocatively, knowing I'd get a testy response given that he'd already made his feelings clear, I asked how he felt towards his former driver, Nelson Piquet Jr. He turned to me with half a smile: 'The fact that we are taking him to court for blackmail, I think so is pretty much what is my feeling.' He could be incredibly charismatic when he wanted to be.

However, Flavio's charm didn't work when it came to the FIA's lawyers. At the hearing a week after Monza, having initially denied the whole plot, Symonds admitted that he had helped to fix the race. As a result, he was banned from all FIA-sanctioned motorsport for five years. I had believed Symonds when he protested his innocence in Monza. He had been a regular contact of mine in the pits, and I trusted him, so when the truth came out, I had a keen sense of having been deceived. Briatore denied any involvement, but this was contradicted by Piquet, and by the testimony of a Renault engineer who alleged that Flavio knew about the plan, even if Flavio claimed it wasn't his idea and that he hadn't been involved in its execution. Such was Flavio's avowed lack of interest in anything technical such as race strategy, this seemed entirely plausible to the rest of the paddock. The engineer had been guaranteed anonymity in return for his evidence so was known, dramatically, as 'Witness X' in the FIA hearing. The FIA ultimately judged that Flavio was indeed involved and banned him from motorsport for life.

Outside the hearing there was a media scrum. Mosley was challenged on the harshness of Briatore's sentence – the flamboyant, charismatic F1 showman, whose world had collapsed. 'It's sad to see a career end like that, but what else could we do?' he said. Max clarified that the FIA had not fined Renault as a team, nor kicked them out of F1 altogether. Despite his distaste for the power of manufacturers, Mosley knew that Renault's future participation was balanced on a knife-edge, and he didn't want to push Carlos Ghosn into shutting down the team.

The only real punishment Renault suffered was from its own title sponsor, the Dutch bank ING, whose management pulled their sponsorship and logos from the car, feeling they couldn't be seen to be associated with a team that had cheated.

With Renault understandably in scandal-management mode, Briatore and Symonds were relieved of their positions. The new team principal was a man who wasn't a fan of the limelight at the time, and possibly as a result of the events he steered the team through, hasn't been keen on it ever since: the likeable Northern Irish engineer Bob Bell. On his first weekend in the job, he was thrust in front of our BBC cameras. He looked like he would have rather been anywhere else, so I tried to put my question as nicely as possible, but there was no way around it:

'So Bob, your team has been embroiled in a race-fixing scandal, you've lost your team principal and your engineering director, and your major title sponsor has walked out on you. How's your first week as team boss going?'

He answered with good humour: 'It can't get worse, can it! I can only look good in a situation like this whatever I do, so we'll deal with it.'

For a man who never wanted to be front and centre, being made team boss at such a moment was a poisoned chalice and one he dealt with admirably. In time he left Renault to join Mercedes, where he would play a key part in Lewis Hamilton's success. He later returned to Renault's Enstone base for another stint with the team before moving to Aston Martin in 2024.

Looking back at the roots of Crash-gate with the benefit of hindsight, it's clear that Pat Symonds was under immense pressure. He felt responsible for the hundreds of employees at the factory and feared that they could all lose their jobs. I'm not even sure Symonds would have considered what he orchestrated as such a terrible crime – Piquet wasn't risking his life in a slow-speed crash, and it was unlikely any marshals or spectators would have been hurt. Maybe he just saw it, to quote himself, as 'a bit of gamesmanship'. Maybe he succumbed to that pressure and hoped at least that it would be a victimless crime.

But it wasn't victimless: in the end the victims were two Brazilian racing drivers. Felipe Massa's Singapore race fell apart when, in the rush for pit stops under the Piquet safety car, a Ferrari mechanic gave Massa the green light to go while his fuel hose was still attached. Petrol sprayed everywhere from the severed fuel line, and Massa lost a minute and a half as the mechanics sprinted down to the end of the pit lane to retrieve the hose. He finished that race a lowly 13th, losing the points that would have won him the title.

Years later, feeling so keenly that Singapore had contributed to his championship loss, Massa went on to file a lawsuit against the FIA for damages.

The other victim was the man who crashed. Nelson Piquet Jr was granted immunity from sanction by the FIA in return for his evidence, leaving him free to take another drive in F1 if he could find anyone who wanted to employ him. He could not and his F1 career was over, although he did continue to race in other categories, winning the inaugural Formula E title in 2015.

Briatore took the FIA to court in France, who ruled that the FIA and its World Motor Sport Council lacked the authority to impose Briatore's lifetime ban. The penalty was overturned and he was awarded €15,000 in damages. In the aftermath Briatore remained on the fringes of F1 as a sponsor finder, but in 2024 then Renault Group CEO Luca De Meo brought him back to Enstone in the role of 'executive advisor'. Meanwhile, the previous Renault boss Carlos Ghosn had a colourful fate. In 2018 he was arrested in Japan, charged with various crimes that centred on alleged improper financial conduct. Ghosn denied the charges, and was held under house arrest in Tokyo while awaiting a trial he felt would not be fair. He thus arranged to be smuggled out of Japan in a musical-equipment packing case via two private jets to his ultimate destination of Lebanon, his home country and one that has no extradition treaty with Japan.

Symonds also reached a settlement with the FIA which allowed him to return to F1, initially with the British team Manor. After a subsequent stint with Williams he was hired by the Formula 1 organization as its technical director, and in January 2025 he started work as a consultant to the new Cadillac F1 team. Along the way he also spent a year on our team at Sky Sports as an expert analyst. And yes, it was awkward between him and me . . .

And the man who benefited most from that controversial evening in Singapore? How much Fernando Alonso really knew about Crash-gate remains a mystery to this day. If anything, it just adds to the mystique that has epitomized his F1 career.

Chapter 13

Brains and Brawn

Lewis Hamilton's spectacular achievement on the last lap of the 2008 F1 season was the perfect way for ITV Sport to go out on a high. The financial crisis and slump in television advertising forced the broadcaster to make some hard decisions, and it opted to end its F1 coverage. The last programme we made for ITV ended with a music piece that pre-dated the American show *Lip Sync Battle* by a good seven years, albeit without the quality. Essentially it was the ITV presenters 'singing' along to 'Welcome to the Black Parade' by My Chemical Romance, the chorus of which began with the words 'We carry on'. True for F1, but sadly not for ITV, and most of the production team I had worked with for the past 12 years.

The producer responsible for the segment was Chris Holding, who did particularly well in persuading F1 drivers and team bosses to join in. 'Once a few start doing it, they'll all want to do it,' he told me while he was in the process of setting it up, basically inventing TikTok at that very moment. I roped in BMW Sauber team boss Mario Theissen, who took in good humour my wearing a false moustache next to his very bushy, real one. It was excruciating viewing, but at least Hamilton, Sebastian Vettel, David Coulthard, Jenson Button, Rubens Barrichello, Alex Wurz and Murray Walker all took part, so I was in good company.

With ITV leaving, Bernie Ecclestone moved swiftly to offer the UK F1 broadcast rights back to the BBC. Given their licence-fee funding method, the corporation did not run adverts in the Grand Prix programmes, so it was a welcome return to uninterrupted coverage for our audience. The BBC Sport production team was led by Mark Wilkin, who had previously produced the coverage with Murray Walker and James Hunt in the years before ITV's stint. Jake Humphrey brought a fresh style to the presentation line-up alongside what grew to become a memorable punditry double-act between David Coulthard and Eddie Jordan. Martin Brundle was commentating, with Jonathan Legard, who had come in from BBC Radio 5 Live, and Lee McKenzie joined as my fellow pit lane reporter. Lee had a more personal connection to F1 in that her father Bob was the Grand Prix correspondent for British tabloid the *Daily Express* for many years, although he gained greater fame running around Silverstone on the 2005 British GP weekend wearing only body paint and a sporran after losing a bet with Ron Dennis!

As pit lane counterparts Lee and I developed an effective and enjoyable professional relationship that has endured to this day, despite the fact that we no longer work for the same broadcaster. In our stand-ups to camera we would follow up on incidents and stories that had come up over the weekend. Lee would introduce the topic, then I would get over-excited and go off into the deep end about why I thought it was the biggest story ever, bringing in loads of details. Lee would courteously hear me out, before turning back to camera to say, 'Yes, well, thank you very much for that, Ted, but just to confirm the main point, which is that Sebastian Vettel has avoided a grid penalty, and will start on pole position after all . . .'

For the first time since the end of 1996, F1 was back on the BBC and motorsport was about to gift it a 'phoenix from the ashes' story filled with grit and determination starring a car that came close to being mothballed: 2009 will forever be remembered as the Brawn GP year. The team's story began at the end of 2006 when Ferrari technical director Ross Brawn decided to take some long-craved time out. He'd been at the coalface at Maranello for a decade, and the pressure of the workload and competition had taken a toll. With Michael Schumacher retiring he wanted to spend more time with his wife Jean and their family, and to pursue the rather more relaxing hobby of fly fishing.

At the same time as Ross was thigh deep in waders in a chilly Scottish stream, the Honda Motor Company was growing frustrated with the lack of progress of its F1 team. Since graduation from engine suppliers to team owners with the purchase of the BAR team in 2005, Honda had invested heavily in F1, but only achieved a single victory with Jenson Button in the rain-affected 2006 Hungarian GP. In 2007 Honda's car was hopelessly uncompetitive. The unusual 'Earth Dreams' livery reflected the fact that there were few sponsors, and that Honda was underwriting the whole thing for little reward. Team boss Nick Fry couldn't fail to be aware that things weren't progressing under his leadership. A pragmatist, looking for a solution, he went on a charm offensive to Brawn, phoning him every week or so, sounding him out as to whether he felt he'd caught enough fish, attempting to lure him back to F1 as Honda team principal, leaving Fry to concentrate on the commercial side and general administration.

It worked. Along with a very competitive salary, Fry convinced Brawn to join Honda because it was an interesting challenge and as team principal (a position Brawn had not been offered at Ferrari), his knowledge, methods and calm presence could have a transformative impact. Honda had finished a lowly eighth in the 2007 constructors' championship, and Brawn arrived too late to have any impact on the 2008 car. If anything, the RA108 was even less competitive than its predecessor, and it left Button and his teammate Rubens Barrichello struggling to get into the points.

What Ross could do straight away, however, was to take steps to improve the Honda F1 organization, and make big strategic decisions. The most significant was his call to halt any development of the hopeless 2008 car and focus instead on the following year's model and the opportunity that came with a blank sheet of paper, thanks to new aerodynamic rules that were coming in for that season.

This was the era before FIA aerodynamic testing restrictions put a limit on how much wind-tunnel running and Computational Fluid Dynamics research teams could do. The sky was the limit, and as 2008 progressed the Honda team worked on its future car in three wind tunnels, including one in Japan, with the aim of hitting the ground running in 2009. But at the end of November 2008 came a hammer blow. Fry and Brawn were called to a hurriedly arranged meeting with Honda executives and were stunned to be told that the company was pulling out of F1 immediately, and that the team was to be shut down. The decision was a reflection of the same global financial crisis that would also lead to both Toyota and BMW leaving the sport at the end of 2009.

Fry and Brawn managed to stall the closure of the team and persuaded Honda to keep things ticking over while they sought a buyer. Some serious players were interested, including Richard Branson's Virgin Group, Prodrive boss David Richards, and even Bernie Ecclestone. However, after careful consideration none was deemed to be the right fit. Eventually Honda resigned themselves to the fact that keeping the team alive via a Brawn and Fry management buyout would avoid the near $100 million cost of closing the doors and making everyone redundant. It would also be less embarrassing and more ethically correct, a highly important consideration for the Japanese company. As part of the deal, Honda would provide modest funding to keep the race team going.

It was a sensible resolution. Even though Fry and Brawn were taking on all the financial responsibility, they were still buying an F1 team for practically nothing. Well, not quite nothing. At the formal meeting to sign the ownership documents, Brawn reached into his pocket and pulled out a one-pound coin, which he gave to Honda's managing officer Hiroshi Oshima. Ross Brawn bought an F1 team for a pound! It was a gesture, but one appreciated by Oshima, who still has the coin. How Honda's financial people feel about that looking back on the deal is another matter. In 2010 Brawn sold the team to Mercedes for around £150 million, and the Mercedes F1 team is now worth an estimated £3 billion.

One of the first things to sort out was a team name. Fry had been keen on Pure Racing, but in the end it was decided to name the team after the boss, much like Ferrari, Williams or McLaren. And thus Brawn Grand Prix was born. There was one sting in the tail related to the Honda exit. The company may have sold the team to Ross Brawn, but it would not let him use its engine. The 2009 Honda V8 was expected to be an improvement on the woeful 2008 powerplant, but it was still likely to be less competitive than its rivals. Not being tied to using it was in one sense a bonus. However, Brawn now had to find, and pay for, an alternative engine supply.

As a rule, rival teams would be loath to help a fellow competitor, but there was a bigger political picture at play in F1 at the time that suspended the normal operating rules of the Piranha Club (the nickname by which the F1 team principals were known, alluding to the voraciousness with which they wanted to destroy each other). 2008 was a year of rare unity, in which the teams had allied to form the Formula One Teams Association, or FOTA. In essence they believed that as the circus performers, they were being undervalued and should be paid more via their commercial agreements with the ringmaster, Bernie Ecclestone. Teams were thus unusually open to the notion of assisting Brawn in order to maintain a show of unity, and to keep his two cars on the grid.

Ferrari boss Luca di Montezemolo offered an engine, but when Brawn discovered that what was on offer was the previous year's power unit, he declined. Then, a lifeline, when Mercedes put themselves forward as a possible partner but with a catch. McLaren was the works-backed Mercedes team, and until 2008 had exclusive use of its V8 engine. For 2009 a supply had been granted to Force India, with McLaren's permission, as part of a bigger technical deal. Now Brawn became an option as well – but only if McLaren would allow it to happen.

McLaren boss Martin Whitmarsh was very active in FOTA, and he understood that if he used his veto to stop Mercedes from supplying Brawn, it would undermine the unity of the teams, and support Ecclestone's view that they were only out for themselves. Logically Whitmarsh has to have assumed that Brawn wouldn't be a threat, given that the team was scrabbling around just to stay alive, and it would also have to convert its new car to accept a Mercedes V8 instead of the Honda engine it was designed for. With this in mind, Whitmarsh duly waived McLaren's veto and gave permission for Mercedes to supply Brawn with their engine. My suspicion has always been that had Ron Dennis still been in charge of McLaren, he would not have been quite so willing to let Brawn have the Mercedes powerplant. However, he had recently handed day-to-day decision making to Whitmarsh and couldn't overrule him on his first big call. It would turn out to be a sliding-doors moment in F1 history.

While all this was going on, Brawn's engineers continued to work on the new car. The changes to the technical regulations from 2008 to 2009 had several intentions. There was a thought that the aerodynamic complexity of the cars had spiralled out of control, and in order to cut down on some of what was seen as wasteful spending on ever more weird and wonderful winglets, and to continue the FIA's constant quest to keep a cap on speeds, the 2009 cars were made to look a lot simpler than their immediate predecessors. It was also hoped that the reduction in downforce would make the cars less sensitive to turbulence, resulting in closer racing. The new regulations made the cars look quite strange, with wider and lower front wings paired with higher and narrower rear wings. Even now, the rear wings of those cars look out of proportion, and at the 2009 car launches I certainly thought so. However, after the first half hour of watching them on track, we got used to the new, lower downforce bodywork.

Over the winter most teams estimated the downforce loss would make their cars a few seconds per lap slower. However, some months before they found themselves fighting for their survival, Honda's engineers had found a loophole in the rules that would enable them to claw back much of the downforce lost in the rule change. During that extensive aero research conducted in 2008 an aerodynamicist called Masayuki Minagawa, who had previously worked with Honda's second team Super Aguri, noticed that the wording of one of the rules referred to bodywork that is 'visible from beneath the car'. His literal interpretation was that this didn't apply to bodywork that wasn't visible from beneath the car. It gave Minagawa the scope to design a section of the floor ahead of the diffuser which had an increased surface area. It allowed more air into the actual diffuser, expanding and accelerating the all-important airflow. This in turn generated a high level of rear downforce – as much as 40 per cent of the total downforce of the car.

When we first started hearing about this design concept, rival teams called it a 'double deck' diffuser, because the two elements sat on top of each other, and worked together to generate extra downforce. The design later became known simply as the 'double diffuser'. However, it was almost done away with before it even raced at the suggestion of none other than Ross Brawn himself.

During 2008 Brawn had seen Minagawa's double diffuser idea, and had recognized that it was so effective that it was likely to claw back all the downforce that the new regulations were intended to cut. In the spirit of team unity fostered by FOTA, Ross brought up the issue in the FIA's Technical Working Group, where he asked the other teams if they thought that they should tighten up the rules. The consensus was 'no', so the loophole was not closed off. Brawn's conscience was clear. It was actually Renault's Pat Symonds who was the crucial player in declining to shut down the loophole, because his team was also working on a double diffuser. They had run their design past the FIA's Charlie Whiting, who gave an opinion that it was not legal, so the Renault engineers hadn't pursued it. However, Charlie's opinion was just that, an opinion. Brawn was more confident, and happy, should the need arise, to take the design to a higher legal authority. As a result Honda/ Brawn and two other teams, namely Toyota and Williams, all incorporated a double diffuser on their 2009 models.

After a fraught winter of survival, Brawn GP wasn't ready to run at the first winter test because the team was still modifying the chassis and gearbox in order to accommodate the Mercedes engine. The Brawn BGP001 eventually turned its first wheel in a private shakedown at a misty Silverstone. Jenson Button wore his red, white and blue British Union Flag crash helmet which clashed horribly with the bright yellow of Brawn's new team identity. He'd have a new helmet in those colours ready for the first race – but that was the least of his concerns.

A few days later the Brawn team trucks rolled into the Circuit de Catalunya for the second pre-season test. On the first morning we walked down to the last garage to see all the familiar faces we'd known from Honda – Ross Brawn was in the pit lane, while Button, Rubens Barrichello and engineers Andrew Shovlin and Jock Clear hung out in the garage. They were wearing unbranded black clothing, as there was no proper team gear yet. McLaren's Dave Ryan would have been horrified.

The team took some time to get the new car ready to run on that first morning, but once on track, it flew. Button's first flying lap was six-tenths of a second quicker than anyone else. He came back into the pits after a few laps of even better times and Shovlin said, 'No need to push, you're already a second quicker than everybody.' A disbelieving Jenson replied, 'I wasn't pushing!'

The car's performance wasn't just down to its double diffuser. When I first saw the Brawn chassis up close, it instantly struck me that it was much more developed for the new rules than any of the others up and down the pit lane. It had obviously had more time spent on it, and the detail was very impressive – the front wing endplates, the neat undercut below the sidepods. After those stunning first laps, Brawn subsequently focused on running heavy fuel loads in order to mask the car's pace and not attract too much attention. It thus appeared from the outside that those early fast laps were 'glory runs' set with a light car. This made it easier for Brawn's rivals to be sceptical about the BGP001's ultimate pace. We had a lot of people, from Christian Horner at Red Bull to Felipe Massa at Ferrari, wondering aloud in interviews whether Brawn had taken a lot of fuel out, or had not run any of the ballast to make the weight limit that most cars did, in order to be as quick as they were on those first laps.

There's a long history of F1 teams doing this in order to attract sponsors, but even though the car had not a single sponsor logo on it, to me it just didn't seem Ross Brawn's style to mess around with underweight cars in order to set deceptive lap times. Nevertheless, it is to my eternal regret that I didn't buy into the story of Brawn's stellar testing form, and remained sceptical of the hype that the team was going to dominate the first race in Australia. Yes, I could see the car was the most developed of them all, and the double diffuser was so obviously different that I was willing to believe that Brawn would be a significant player in the mix. But I also thought there would be enough competition from the likes of Red Bull, Ferrari and BMW Sauber to make a close race of it in Melbourne.

In fact, Button would score a dominant victory from pole position, although it took Sebastian Vettel running into the side of Robert Kubica when battling for second place to allow Barrichello to recover from a bad start and come through to make it a Brawn GP one-two finish in the team's first race. I headed down to the team's garage to find Jenson's father John Button leading the celebrations in the pink shirt he always wore on race day. The mechanics and engineers looked ecstatic at what they'd achieved. Lee McKenzie grabbed a tearful Ross Brawn as he came off the podium with the winning constructors' trophy in his arms, and we analysed the race on the new BBC F1 'Forum' show until darkness descended on Albert Park.

In the emotion of that Sunday evening it was easy to forget that the result was still provisional pending a session of the FIA Court of Appeal. Ferrari, Red Bull, Renault and BMW Sauber had jointly protested the double diffusers on the Brawn, Williams and Toyota cars at the start of the weekend, but their protest had been rejected. Their next step was to escalate the case to the more formally legal surroundings of the appeal court. The teams that didn't join the protest were Brawn's fellow Mercedes-powered squads, McLaren and Force India – and Toro Rosso, because Red Bull was already represented.

FIA Court of Appeal sessions are often dreary affairs, with more lawyers than you'd ever want to have in a confined space quibbling over semantics. Not this one. It was a cracker, the star of the show being Nigel Tozzi QC, widely regarded as one of the preeminent barristers on the planet and engaged, just like he was in the Spy-gate hearings, by Ferrari. Tozzi (who had worked with Brawn when he was at the Scuderia) absolutely laid into Ross, accusing him of being 'a person of supreme arrogance', before criticizing Charlie Whiting for 'not understanding the point' and 'getting it wrong'. Whiting had seen this all before and, as I recall, knowing it was all bluster, found it quite entertaining.

Unfortunately for Tozzi and the other appellants the argument that the double diffuser was 'inconsistent with the spirit of the rules and the efforts of the FIA to facilitate overtaking' was completely torpedoed when Brawn proved that Renault had tried to do exactly the same with their double diffuser concept, so were hardly in a position to object. Pat Symonds must have wished he'd never asked Charlie Whiting's opinion!

Not only was that point dismissed, but all the others were too – including the legal debate over when a fully enclosed hole is classified as a hole, and when it's just a space created by discontinuations of surfaces, in which case it's not a hole! The intention of the rule was supposed to result in a flat, unimpeded area with no holes in it, but ultimately Brawn's (or Minagawa's) interpretation of the regulation as written was deemed legally correct. The appeal was dismissed, the Australian GP results stood, and the advantage swung firmly in Brawn's favour.

But it wasn't all plain sailing. Button's win in Melbourne was followed by victories in Malaysia, Bahrain, Spain, Monaco and Turkey, after which point (in early June in a season that would stretch to November) Jenson would not win again. There were three reasons for this. Firstly, people. The champagne of Melbourne had barely dried when Fry and Brawn had to make a painful round of redundancies. The team was downsized from 750 to 400 employees as a cost-saving exercise, so Brawn GP was forced to operate with a smaller staff than its rivals.

Secondly, the development budget. The double diffuser had given Brawn a colossal early-season advantage, but there was no money to develop the car and maintain that sort of form. In an interview we did at one of the summer races, Button's engineer Andrew Shovlin was one of the first to publicly admit that, because Brawn GP didn't have much money, there was no development of the car happening. It was clear that the team was really struggling.

Thirdly, the opposition caught up. McLaren was the first rival to come out with its own double diffuser at the third race in Shanghai, and while Adrian Newey had a tougher challenge to incorporate it on to the Red Bull because of earlier gearbox and suspension design choices, the RB5 became increasingly competitive.

For these reasons, and some others, Jenson Button had something of a mid-season wobble. He became frustrated, made mistakes and generally lost the cool, calm and collected demeanour he'd had at the start of the season. The pressure was building – Vettel was catching up, as was Jenson's own teammate Barrichello, who won in Valencia.

After a mid-grid qualifying result in Belgium, Jenson tangled with Romain Grosjean on the first lap, and spun off into the barrier. That non-score looked like it might be a serious blow for his world championship chances, but luckily for Jenson it was Kimi Räikkönen who won at Spa, rather than Vettel or Barrichello. A crucial fifth place in Singapore and an eighth in

Japan kept the points coming. And then came Brazil, Barrichello's home event.

Button took a week off after the previous race in Japan. He had a few days in Tokyo with his partner Jessica Michibata before flying to Hawaii for a mini break. He spent the time eating, sleeping and training, knowing he'd need every bit of energy he had for the Brazilian GP weekend. It's quite difficult to get from Hawaii to South America, and for Jenson, it involved two whole days of air travel via Los Angeles and Miami to São Paulo. He had always enjoyed the Brazilian event, ever since his first race there with Williams in 2000. He loved the Interlagos circuit, the fans, the food and, like the rest of us, when the weekend's duties allowed, a caipirinha, the notoriously strong Brazilian cocktail. John Button was also a keen visitor, and I would sometimes find him taking a break at the back of the Interlagos paddock where the balcony looks out to the circuit below. He'd always joke about how it reminded him of Brands Hatch.

Unfortunately, the Brazilian reception extended to the Button clan in 2009 was not a warm-hearted one. Even with the emergence of Felipe Massa, who had come so close to winning the championship the year before, Brazilian fans still had a soft spot for the seasoned campaigner Rubens Barrichello. They understood that this was probably the last chance that he would have to win the world championship, and Barrichello had stoked the fire a few races earlier by publicly criticizing Brawn GP's race strategy, with the obvious subtext that he felt the British team with its British boss preferred their British driver to win, not the Brazilian.

After the black-cat prank on Lewis Hamilton hadn't proved effective, the same local TV show that pulled the stunt on Lewis the year before turned up and doubled down on Jenson. The prank occurred on the Tuesday evening before the race, when Button, his dad, his trainer Mikey 'Muscles' Collier and manager Richard Goddard drove to the Fogo de Chão steak restaurant, a venue beloved of the F1 paddock and a favourite stop for drivers. The two TV presenters, one wearing a chequered flag suit and the other a plastic wig and false teeth, set up a large A-frame ladder over the restaurant entrance, obliging Jenson to walk underneath if he wanted to get to his steak dinner.

Unlike crossing the path of a black cat, walking under ladders *is* usually considered bad luck both in the UK and in Brazil, so the pranksters thought their plan foolproof. Except they initially mis-identified Collier as being Jenson, while the real man slipped round the back of them, therefore deftly avoiding the on-camera moment they'd hoped for. Jenson still had to walk under the ladder, but declared in his easy way that he wasn't superstitious, so the plan was wasted. Nothing if not determined, the jokers followed Button back to his hotel, where they repeated the trick, this time in the car park, where Jenson's whole car had to pass under the ladder once again. Adding insult to injury they then set off loud fireworks outside his hotel in an effort to disrupt his sleep. It made for a completely outrageous piece of TV, and Jenson did well not to react to it.

The weather in São Paulo is notoriously changeable, and a huge rain shower made for difficult conditions for everyone in qualifying. Neither Hamilton nor Vettel made it out of Q1, Vettel's misfortune being particularly good news for Jenson's championship chances. Then in Q2 a season-long characteristic of the Brawn BGP001 would prove troublesome once again, but at a crucial moment.

The car had been engineered with the beneficial characteristic of being easy on its tyres – it tended not to work them too hard or cause them to overheat and lose grip. In some of the hot, early races, and in Valencia in the middle of the summer, this was a positive attribute. However, at cooler track temperatures, such as those found at Silverstone, it meant that the car couldn't generate enough temperature to get the tyres in a sweet spot delivering grip to the driver. Couple that with Jenson's characteristic driving style, all smooth inputs and gradual steering, and you had a car that at some tracks could be completely different to the machine that had dominated the first third of the season.

One such circumstance was the drying track that presented itself in the second part of qualifying at Interlagos. A key bit of indecision in Shovlin and Button's communications meant that Jenson was left out on his full-wet tyres too long, when he should have pitted for intermediates. Shovlin had reasoned that since the car was generally easy on its tyres, the full-wets would last one more lap. They didn't – Button had no front grip, and the rear tyres overheated.

Explaining Button's sudden nervousness to me at the time, Ross Brawn said he believed that the consequences of making a potential mistake 'became prevalent in his thinking'. Martin Brundle described Button's qualifying problems as 'like a golfer with the yips'. In Brazil, Barrichello was on pole position. Yips or not, Jenson was starting a distant 14th on the grid. Maybe the ladder trick had worked after all.

That Saturday night, keen to avoid whatever hideous prank the Brazilian TV presenters had up their sleeves, Jenson had a quiet dinner with his entourage in the Japanese restaurant at his hotel. He had a beer, checked in with his mum on the phone, and went to bed.

On Sunday morning, something clicked. By the time he got to the track, Jenson was up for the fight, and he knew that he had it in him to get the points he needed to close out the world championship. As he prepared for the race, he told Ross Brawn not to worry, and that he was going to make up for the lowly grid position. He then proceeded to drive like an absolute demon. Later, in his book *Total Competition* (2016), Brawn revealed he had been perplexed by Button's sudden aggressive precision: 'He came from behind, overtook people, he was decisive, he was really impressive. But then I thought, "Why can't you do that every race?"'

The truth was the sudden change in driving style took Button's opponents by surprise. Two overtakes in particular, on Romain Grosjean and Sauber debutant Kamui Kobayashi, were heart-in-the-mouth in terms of how risky they were, and how close to going wrong, and another on Sebastian Buemi, whom Button knew he had to take by surprise, because the Toro Rosso driver had very good straight-line speed and would be hard to pass. It's an often-used phrase, but it was a champion's drive, delivered at just the right time. By the chequered flag he had clawed his way up to fifth. There was one race in the season still to run, but Jenson Button had achieved the points he needed to win the 2009 world championship. By the time we got to him on the pit wall Brawn was in tears – in fact there was hardly a dry eye in the team's garage. Winning the constructors' world championship as well meant so much to the whole team because of what they'd been through together.

On the slowdown lap Button proved that the one thing he didn't have that weekend was a great singing voice as he belted out the chorus to Queen's 'We Are the Champions', slightly off-key. He was so openly ecstatic after the race that it was just as well he'd saved up some energy. The relief at getting the title wrapped up before the last round, the release of stress and pressure, and the sheer joy at not only winning, but doing it in such style were clear to see. Jenson had spent 21 years working on his dream of being world champion, and he had achieved it.

The story of Brawn GP was one of the greatest underdog tales that any sport has ever seen, but it could only last a single season. That sliding-doors moment when Martin Whitmarsh agreed that Mercedes could supply Brawn would come back to haunt him. Keen to have its own modern-day Silver Arrows team, Mercedes bought Brawn GP, renamed it, and began the process of withdrawing its works support from McLaren. The Woking team would not win another title until 2024.

Barrichello left to drive for Williams, and Button joined McLaren, where he had an enjoyable and successful end to his F1 career. He still races in other categories, and works as a pundit at Sky F1, and is every bit the lovely, easy-going, smart, sparky guy you'd think he is. His father John died in 2014, and is still missed and remembered with great affection. If you ever see me wearing a pink shirt on a race day, you'll know who inspired me.

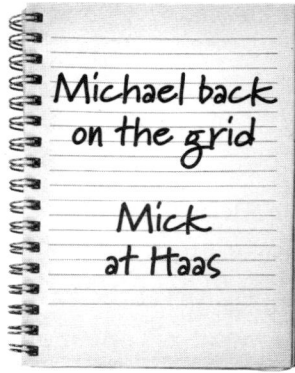

Michael back
on the grid

Mick
at Haas

Chapter 14

Schumacher's Second Coming

Following his retirement at the end of 2006 Michael Schumacher was appointed a 'super advisor' to Ferrari by Jean Todt, keeping him connected to F1 during the 2007 season, but when he turned up in the paddock he always looked uncomfortable. Schumacher would typically watch the practice sessions from the pit wall but then retreat to the hidden racks of telemetry screens for qualifying and the race. His former teammate Felipe Massa was happy to have Michael around and ready to listen to his advice – Michael's replacement Kimi Räikkönen rather less so.

Michael could sense this, and I always observed the two keeping their distance, which just added to Michael's low profile. Whenever I'd pass him in the paddock, he'd give a quick 'howdy' (slightly odd, you might think, from a German speaker, but Michael liked the cowboy persona), but he wouldn't stop to chat. His manager Sabine Kehm would organize the occasional press conference in which Michael would handle questions about whether he was planning a comeback with a good-humoured 'no'. He knew that F1 was constantly evolving and seemed content enough to watch the ongoing rivalry between Alonso, Hamilton, Massa and Räikkönen.

When Schumacher was given the chance to sample the 2007 championship-winning Ferrari at the Barcelona post-season test he showed he'd lost none of his speed, but in general he was becoming more and more interested in two-wheeled motorsport. When he was racing, he would often be seen arriving at F1 paddocks on his custom Harley-Davidson and as part of a Ferrari-Ducati tie-up had even tested a MotoGP motorbike at Mugello, but being so focused on his F1 career, didn't have much more time for anything else. Once freed from his driving contracts, though, the idea of learning a completely new discipline appealed, and he enjoyed the sensations that racing motorbikes provided. He entered the German national Superbike series and was soon lapping within a couple of seconds of the top riders. Given that he was taller and more muscular than most motorbike riders tend to be and hadn't raced bikes all his life in the way they had, this was a seriously impressive performance.

Stories about his bike exploits would reach us in the F1 paddock, but at the beginning of 2009 news came through that he'd had an accident. It was at the Cartagena circuit, a small but well-used test track on Spain's southeast coast. Details were few and far between, but initial reports were that Michael had not been seriously injured.

The F1 season started and we all got wrapped up in the Brawn Grand Prix story. It wasn't until halfway through the year we discovered what that Spanish motorbike crash would mean for the rest of Schumacher's racing career.

At the Hungarian GP a rear suspension spring fell off Rubens Barrichello's Brawn, bounced down the track and hit an unsuspecting Felipe Massa's helmet, taking a sizeable chunk from the top left quadrant. It was a shocking accident. Massa was knocked unconscious, the car went head first into the tyre barrier and he narrowly escaped losing the sight of his left eye. It was impossible for him to race and he was out for the rest of the 2009 season.

It was assumed that Ferrari would call on their established test driver Luca Badoer as a stand-in for Massa, but we soon learned that Ferrari president Luca di Montezemolo and new team principal Stefano Domenicali wanted Michael back. He might nominally be retired, but his skill and experience vastly outclassed that of the journeyman Badoer. He initially said no when Di Montezemolo approached him with the offer, but was persuaded to test the 2007 car again. Nine days after Massa's crash, Michael pulled on the famous red race suit and headed to the Mugello circuit in Tuscany.

'He went into the racing department and was full of enthusiasm, like a kid or a young driver. Then he went to Mugello and did a very good test with the old car,' Di Montezemolo reported. Ironically it was Di Montezemolo himself who had done most to pressurize Michael into the decision to retire in 2006, but now he listened to the instinct that told him Michael was ready for all that ability, desire and motivation to come to the fore once again. For his part Schumacher confirmed that the Mugello test reminded him just how much he had been struggling with life outside F1. 'Although I was a retired race driver, still for a moment, I felt back alive.'

Just a few days after the Mugello test, Sabine Kehm invited us to a press conference with Michael and members of his team at the Intercontinental Hotel in Geneva, not far from his home. We knew we had to cover it, so I jumped on a plane, met up with a BBC News cameraman in Switzerland, and set up ready for the conference. Before it started, we had the big announcement from Ferrari that Badoer would deputize for Massa after all, so the news that Schumacher's Ferrari comeback was not to be didn't come as a surprise.

It was clear as soon as Michael walked into the conference room that this was going to be tough. He looked miserable. Alongside him were his long-time manager Willi Weber, and his doctor of nine years, Johannes Peil, a tall, slim man with a kind face, a tidy moustache and a shock of blonde-grey hair. Schumacher began by saying how frustrated and sad he felt, and how disappointed he was not to be able to help Ferrari by standing in for the injured Massa. He then explained why. It was that motorbike accident in Cartagena, and only now did we discover how seriously he had been injured. He had broken a rib and fractured two vertebrae in the fall. They had since healed, but there was one remaining injury that had not healed, and it was the pain from that physical damage that made driving an F1 car for any length of time impossible.

As Michael sat listening intently and Willi Weber looked out into the room with a slight grimace, Dr Peil explained that there had been a tear in the left-hand side ligament that links the base of the skull to the first vertebra at the top of the neck, named the C1 or Atlas. This tissue around the Atlas played a highly important role in protecting the brain. This had not healed, and it was giving Michael so much pain that he was not able to drive the car. Schumacher admitted that if he'd had only a small pain in the neck, he'd have treated it with painkillers and got on with driving. However, Dr Peil also had concerns that any accident over the last half of the 2009 season could further damage this area of the skull, leading to long-term injury, or even paralysis.

Dr Peil said the tear could take another six or even twelve months to heal, or might never completely heal. Schumacher refused to completely rule out making a return to F1, but it would not be in 2009. Peil confirmed that aside from the neck issue, Michael was in top physical shape. An intensive workout regime had resulted in him losing 4kg, and his arm, shoulder and leg muscles were up to the job. It could have been a highly successful comeback.

Michael was visibly upset that it wasn't going to happen, saying that this was one of the toughest moments of his career. Weber remarked on how fast Schumacher had clicked back into the racing driver mindset.

After an hour and a quarter, the press conference came to a close. Michael thanked everyone for coming at short notice, stood up, had one last look around the room, puckered his lips up into a sad expression, and left by a side door. While my fellow reporters stood up to check additional details with Kehm, and my BBC cameraman arranged to feed the footage back to London, I reflected that while this had been a frustratingly unfulfilled episode in Schumacher's motor racing career, his comment about 'feeling back alive' when driving an F1 car was, for me, the big story. Michael Schumacher was burning to come back – and fight for that elusive eighth world championship.

A few months later the comeback became a reality – but not with Ferrari. Late in 2009 it was announced that Mercedes was buying the Brawn Grand Prix team and turning it into a full works outfit. When Brawn and Mercedes motorsport boss Norbert Haug asked Michael if he would like to join the newly re-named team, it didn't take long for him to say yes. It was a dream signing. Jenson Button and Rubens Barrichello had both moved on. With fellow German Nico Rosberg as teammate in the re-born Silver Arrows, what could go wrong? 'You take the world championship-winning team at the end of '09, you take Mercedes and you take me,' explained Schumacher. 'So you think you've got to be fighting for the Championship.'

Unfortunately, the 2010 car proved not to be competitive. During the team's 2009 season Brawn GP had spent so much effort fire-fighting problems and trying to get both championships over the line that they had fallen behind in development and lost any advantage from the double diffuser as everyone else caught up. By the time Mercedes took over and the team began to build up its resources, the 2010 car trailed far behind the main opposition. It went from the front of the grid to scrapping for the minor points in the space of nine months.

Rosberg demonstrated that he was able to extract more from the Mercedes W1 than Schumacher, out-qualifying and out-racing the master over the course of 2010 and 2011. I had too much respect for Michael to ask him on-air why his comeback was going so badly. In reports I'd mention factors that could explain his performances, while everyone – the Mercedes team and the media – desperately searched for any glimmers of success. There was a brilliant lap in Monaco qualifying in 2012, and the only podium of his comeback in Valencia later that year, but that season would be Schumacher's last. The team poached Lewis Hamilton from McLaren for 2013. It was time for Michael's second farewell, which came in Abu Dhabi. I remember his wife Corinna looking forward to their family future outside F1. 'At last,' she said, 'we have him back.'

It was the right time for Schumacher to retire for good. The mantle of 'best-current-driver' had already been passed on to Fernando Alonso, then Lewis Hamilton and coming through was Sebastian Vettel. Such is the speed at which F1 moves, even a name like Michael Schumacher was quickly consigned to the history books as the new drivers took centre stage.

Michael was free to spend time with his wife and kids, Gina and Mick, the latter of whom was continuing the family business. Mick Schumacher had started karting around the time of Michael's first retirement using his mother's maiden surname of Betsch in an attempt to divert attention and expectation. It didn't work. Everyone inside the karting community knew exactly who Mick's dad was, and anyone outside it would have been able to work it out pretty quickly just by looking at Mick's familiar facial features and consulting the internet. His results were good, and by 2013 Mick Schumacher was finishing in the top three of most karting championships he entered. There was good reason to think that Michael's genes and guidance would shape Mick into a champion of the future.

We had not long wrapped up our coverage of the 2013 F1 season on Sky Sports when, a few days after Christmas, we heard the news that Michael had been in a skiing accident in France while out on the slopes with Mick. We understood that he had hit his head on a rock. He was wearing a helmet, so when he was airlifted from Meribel to hospital in Grenoble initially we had no reason to be alarmed. Later that evening, however, Sabine Kehm sent an email saying that there would be no running updates regarding Michael's health, and a press conference the following day confirmed the seriousness of his brain injury.

Over the next few months my thoughts about Michael's situation crystallized into two strands. Those of us in F1 who had watched him, worked with him, witnessed his triumphs and setbacks, had not the tiniest doubt that here was a man of such physical and mental strength, determination and stamina that he would be as well placed as anybody could be to recover from his injuries. If anyone could get through this, Michael could. One of the greatest champions Formula 1 has ever known, a charismatic and vibrant figure both on track and in the paddock, we sent him every best wish for the fight he had to come.

My second thought was on the cost of fame and stardom in F1. A guy from an ordinary background used his extraordinary gift to excel in his chosen sport. This attracted interest, appreciation and admiration. He had become so well known worldwide that there was an inexhaustible demand for information about him, particularly at this, his toughest moment. I watched the press conferences from the hospital in Grenoble and thought about all the similar setups where I'd seen Michael, sat behind some table, microphones spread out in front of him. Whether it was winning another world championship, giving his side of a controversy, or even letting us in at a vulnerable moment – Michael understood and appreciated that people were interested in him, and he wanted to give back. But now, when he couldn't give back, it felt to me as if people shouldn't be pushing for answers when there was so little to say. The fame and fortune of F1 stardom had a price, and it was being borne with strength and dignity by Michael's family.

When he began turning up regularly at Grand Prix weekends, Mick Schumacher carried himself lightly, seeming happy not to draw too much attention to himself. Signed up by the Ferrari young driver programme, his time in F3 and F2 had followed a pattern, with the first seasons being learning years, before he stepped up his game and won both championships on his second attempt. With F3 and F2 titles under his belt and Ferrari's support behind him, an F1 graduation was inevitable. He tested the Alfa Romeo-branded Sauber in 2020, hoping to get the race drive, but with Kimi Räikkönen and Antonio Giovinazzi (a young driver one step ahead of Mick on the Ferrari ladder) already signed up he was placed instead into the Haas team for 2021, alongside fellow rookie Nikita Mazepin.

I'd seen Mick around the paddock and said hello, but it wasn't until the last weekend of 2020 when he'd been announced as a 2021 Haas driver that we sat down for our first interview. I walked down to the hospitality area with my cameraman Pete Velluet to find Mick already waiting for us, accompanied by the familiar face of Sabine Kehm, now the younger Schumacher's manager and media liaison.

After a quick exchange of pleasantries, the interview began. I asked him how he would be approaching the season, to which he answered that 2021 was his year to learn both the team and the circuits. He was really looking ahead to 2022, when a regulation change intended to make the cars less sensitive to aerodynamic turbulence and therefore easier to race had the potential to shake up the field, and perhaps result in some opportunities for smaller teams like Haas. The interview concluded, and Pete started packing away the camera and tripod. Mick took the opportunity to ask me how I saw the Haas team and its prospects for 2021, and how I thought he should play his hand.

Sabine kept a watchful eye on proceedings. It was only the second or third time I'd ever met Mick, but I felt I had earned enough of Sabine's trust over the years to be comfortable giving an honest opinion. 'You need to be an island,' I said. Mick gave me a funny look, pursing his lips in a way so similar to Michael's it made me smile. 'OK, here's what I think. In Haas, you're joining the smallest team with the smallest resources both in terms of finances and factory. In Guenther Steiner you have a boss who is, essentially, a mad genius. He'll be busy keeping all the plates spinning, keeping Gene Haas happy, the team financially afloat, keeping the drivers out of trouble and hopefully scoring some points with you. You have a fellow rookie alongside, and the whole of Ferrari micro-analysing your every lap. This will create several whirlwinds over the season. You need to embed yourself as strongly as you can, be an island in the middle of it, and not get swept away in the storms to come.'

Whether this was more or less of the wisdom that he'd been hoping for I never knew, but Mick nodded his head and we said our goodbyes, aiming to have another chat when he started driving full-time at the start of 2021. His first full season with Haas was indeed a bumpy ride – there were more accidents than Schumacher and the team could afford. His best result that season was 12th at the Hungarian GP, just behind Daniel Ricciardo and Kimi Räikkönen. True to form, his second season was better, but there were more crashes, and a couple of points finishes weren't enough to save his seat. He was dropped by Haas at the end of the season and also left the Ferrari programme. He was hired as a test and reserve driver by Mercedes, his father's last team, but left at the end of 2024. Meanwhile, he forged a new career in sportscars, competing at Le Mans and other endurance races. In F1, being just an island is not enough.

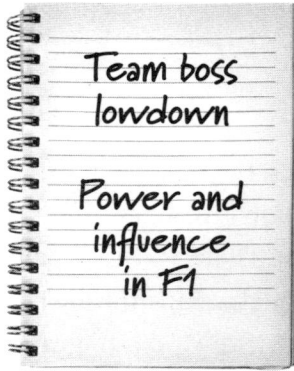

Chapter 15

The Piranha Club

There's a story that when the *Drive to Survive* producers started exploring the world of Formula 1 one of the first bits of advice received from an experienced press officer was, 'If you think the drivers are competitive, wait until you see the team bosses.' It's absolutely accurate – and has a very simple explanation. All team bosses share the qualities of being competitive and single-minded, but history has shown that many of the best were formerly drivers. Enzo Ferrari, Colin Chapman, Ken Tyrrell, Bernie Ecclestone and Frank Williams all came into the sport drawn to the life of the racing driver, but found in time that their true talents lay outside the cockpit. And the fact that they got tantalizingly close to success on the track may have sharpened their motivation even more.

The race drivers are the gladiators, the ones at the heart of the action. Team principals might be responsible for everything else, but the one thing they're not actually able to do is race the cars. They get keyed up watching their drivers fighting it out on track, and if their team hasn't done well, they tend to take it out on everyone else. And when I say everyone, I mean their drivers, other teams' drivers, rival team bosses, engineers, the F1 organization, the FIA and the media. There really isn't anyone that a team boss won't try to outface when the chips are down. And make no mistake, this is not only a ruthless power game between paddock competitors – it's about surviving at the top. If you can steal a rival's driver, lure away their technical director or seduce a sponsor, you not only strengthen your own team, you weaken the opposition. According to F1 lore, it was Ron Dennis who first coined the term Piranha Club, but we'll come on to that later.

The team bosses back in 1997, when I first arrived in the paddock, were a colourful bunch. Ron Dennis seemed calm and science-driven on the surface, yet underneath proved to be highly emotional. Flavio Briatore was (and remains) commercially driven and relatively unemotional. But Frank Williams was the one who made the biggest impression on me. What motivated him were his guiding principles of what was good for his company and his team, while also holding a strong belief in what was right for F1 as a whole.

You're probably familiar with the backstory, how Frank Williams followed his burning ambition to own and run a successful motor-racing team, how it almost broke him financially, how he managed to overcome every obstacle, how he eventually teamed up with talented engineer Patrick Head and finally found the budget with which to create a winning car. Then, in March 1986, came the road accident in France at the wheel of a Ford Sierra that did break him, physically at least. And yet Frank's incredible resilience and determination allowed him to remain in charge of his Williams team and lead it to even greater successes.

I think most F1 fans have a soft spot for Williams Grand Prix Engineering Ltd, to give it its full name. The team won the constructors' world championship nine times between 1980 and 1997 along with seven drivers' titles with Alan Jones, Keke Rosberg, Nelson Piquet, Nigel Mansell, Alain Prost, Damon Hill and Jacques Villeneuve. Those numbers would surely have been even more impressive had Ayrton Senna not lost his life in a Williams at Imola's Tamburello corner in 1994.

I was always aware of this history any time I asked Williams for an interview, or talked to him in the team motorhome on some background enquiry. For all his achievements, Frank was an incredibly humble man, and it's a mark of the respect in which he is still held that the current owners of the team have never once considered changing the name above the factory door.

Sadly the same wasn't true for Eddie Jordan's team when he sold up to a Russian-born Canadian Steel magnate called Alex Shnaider, who only retained the Jordan name for the 2005 season, before re-naming it Midland F1 for 2006. I went to the final Jordan launch in Moscow's Red Square, a very curious occasion with lots of talk about wealthy Russian sponsors re-invigorating the team.

Jordan himself was another true racer. He'd actually been a relatively successful F3 driver before a big crash at Mallory Park turned him towards team ownership. Some people who dealt with Eddie thought he was primarily interested in money, but that was only because he never had any. Despite his many successes, Jordan Grand Prix was never awash with cash, and Eddie had to make what money he had work hard. But he was an ex-racer, and more than anything he loved the thrill of the sport and the drivers who raced in it. In 1991, having been responsible for giving Michael

Schumacher his Grand Prix debut at Spa, Jordan was then ambushed by Flavio Briatore and Tom Walkinshaw from the bigger Benetton team. They had approached Schumacher to offer him a drive and, believing their car offered him a better chance of winning, Michael wanted to join Benetton. All Jordan had was a piece of paper signed by the German youngster saying he intended to sign 'a contract'. It was not an actual Jordan race contract, so legally, Eddie lost the fight to keep him. It was at this point Dennis is said to have welcomed Eddie to 'The Piranha Club'. The qualities of the South American fish to which he was referring were its powerful bite, sharp teeth and appetite to consume anything. Jordan was as canny an operator as anyone, having begged and borrowed his way from Dublin bank clerk to F1 team owner. But even he was astonished by the audacity of his fellow team principals in their willingness to screw over their rivals. Dennis knew what he was talking about – he'd savaged more than a few competitors himself.

The transfer was good for Benetton, good for Michael, and had met with the approval of Bernie Ecclestone, F1's benevolent dictator, at that time the owner of the lake in which the piranhas swam. Despite the fact that Ecclestone had sanctioned the swiping of Schumacher from under Jordan's nose, he and Eddie understood each other well, and were very close. In Eddie's years as a pundit at the BBC it was Ecclestone who was the source of many a good tip and inside story.

Jordan's F1 team would have been worth more than he got from Shnaider had he been able to keep it going for a few more years, but money was tight back then. He went on to enjoy his retirement and family life with his wife Marie and their children and many grandkids with all the zest that he put into his many years in F1. And to complain about trivial or petty things is something Eddie Jordan never did, just as he didn't have time for those who envied his success. He had a tattoo on his wrist – his sons Kyle and Zak have it too – the three letters FTB. It stands for EJ's favourite Irish motto: Feck the Begrudgers.

It's a pity that we didn't have *Drive to Survive* 30 years ago – cameras going behind the scenes with the likes of Eddie Jordan, Frank Williams, Ron Dennis, Tom Walkinshaw and Jean Todt would have been quite something. More recently, Flavio Briatore's surprise return to the paddock with the Alpine team in 2024 meant that we did get to see some of him in Season 7, and it was no surprise that he was immediately cast as a pantomime villain. The series has made unlikely TV stars out of the current generation of team bosses, and has highlighted just how highly motivated they are to beat each other, not least when documenting the intense rivalry between Toto Wolff and Christian Horner.

Born in 1973, Christian Horner embarked on a career as a racing driver, progressing through karting and winning a few races in British F3. In 1997, with help from his father, he opted to set up his own team for the step up to the next category, Formula 3000. Rather than use the family name he called the team Arden, after the area and forest made famous by Shakespeare, near his family home in Warwickshire. In need of a race transporter, he bought one that was surplus to requirements at Red Bull advisor Helmut Marko's RSM team. Horner is said to have gained the Austrian's trust by paying for the truck, sight unseen, with only Marko's word that it would actually be delivered. For his part, Marko may well have taken note. For a young driver to establish their own F3000 team, buy the cars and employ mechanics and engineers to run them was rare, and hinted at greater ambitions.

Over his two unspectacular years in the category Horner's best achievement was a solitary sixth place in a field that included the likes of future F1 drivers Juan Pablo Montoya, Ricardo Zonta and Nick Heidfeld. He eventually concluded that he wasn't going to make it in his own right – reflecting years later on a podcast that it had been watching the way Montoya committed to high-speed corners that had made him realize he simply didn't have the racing bravery that the best of his rivals displayed. At the age of just 25 Christian hung up his racing helmet to focus instead on running Arden. Over the next few years the team developed into a well-respected and professional outfit, winning the drivers' F3000 title in 2003 with Bjorn Wirdheim. Clearly ambitious, Horner also became the representative of the F3000 team bosses in the F1 paddock, dealing with Bernie Ecclestone, who would become a friend and mentor. He also stayed close to Helmut Marko, who was becoming an increasingly powerful player in the sport as the right-hand man of Red Bull co-owner Dietrich Mateschitz. When Marko was looking for an F3000 seat for his protégé Tonio Liuzzi for the 2004 season, he went to Horner and Arden. It paid off for all parties, as the Italian dominated the championship, and the relationship between the team boss and the sponsor bloomed.

I'd often see Christian Horner in the F1 paddock when Arden had won an F3000 race, or he was going about some piece of business. It was in a passport queue at Milan's Linate Airport that he let me in on a little secret, hinting broadly that we might be seeing more of each other. As we got talking, Christian explained that he was keen to move into F1. He'd been in talks with Red Bull to be part of a takeover of Jordan, but the Austrian company had another interesting option in the Jaguar team currently delivering mediocre results under the ownership of the Ford Motor Company. 'So you might be interviewing me in the paddock next year,' concluded Horner, as I reached the front of the passport queue. 'Right, OK then,' I replied, in a 'Good luck with that, Christian' kind of way. I proceeded to do absolutely nothing with this little well-sourced news nugget. Several months later, to the surprise of everyone except me, it was announced that Red Bull, the Austrian fizzy drinks and marketing company, had bought the Jaguar team – and had employed the 31-year-old Horner as team principal.

Red Bull Racing's statistics under Horner's leadership were impressive: 8 drivers' and 6 constructors' world championships, 107 pole positions and 124 wins, but even this track record wasn't enough to convince Red Bull that he was the right man for the future. Just three days after the 2025 British Grand Prix, he was sacked by the Red Bull parent company in Austria. We've always sparred with one another, possibly because he is only four months older than me, and I see him as a contemporary, despite the fact that we're very different people. I think Christian sees the TV side of F1 as a game. He was intensively media trained before he started as team principal at Red Bull and developed into a savvy interviewee who would just about give you a straight answer if you asked the right question, much like a determined batsman dealing with a tricky delivery from a fast bowler. He made headlines of his own, and it remains to be seen how his next move in F1's Piranha Club works out.

Toto Wolff does things differently. Unlike Christian Horner, Toto owns a third of the team he leads, the rest made up by Mercedes-Benz and Ineos, Jim Ratcliffe's chemicals-to-automotive conglomerate. Wolff came into Formula 1 on the business side, but there was racing in his past too, even if his lanky frame made it difficult to squeeze into single-seaters. He won some races in GT cars before an accident gave him cause to think about his attitude to risk. It happened at the unforgiving Nürburgring in 2009, when he had an idea that he could break the lap record of the Nordschleife in a sporty Porsche with just a foot full of throttle and a head full of belief. A blown tyre put Wolff into the barriers at 180mph, and the 27G crash gave him concussion and a trip to hospital. His career behind a desk has been much more successful. He presided over eight constructors' world championships between 2014 and 2021, although when I first met him, he was still figuring out how to manoeuvre himself into the top job at Mercedes.

We were coming to the end of our first year of F1 coverage on Sky Sports. Sebastian Vettel had signed off 2012 with his third world title, and the circus performers had packed their bags and returned home for the winter. As we still had an active F1 channel to fill with content, we tended to leap at every opportunity – and one such was an invitation from Red Bull. The company was organizing one of its typically attention-grabbing events where extreme sports athletes raced powerful snowmobiles around an old Austrian village. Somewhere along the line of communication we were made aware that Vettel might be attending, and that there could be an opportunity for a fireside chat with him about becoming a three-time world champion. Initially it seemed unlikely that Vettel would interrupt his own winter break by making an appearance, but with each successive conversation with every producer down the line, confidence somehow grew that he would show up. So two camera crews and a sound operator were booked, and I was put forward to go, along with producer Alex Rodger, for no greater reason than we could both ski.

A flight, two hire cars and five hotel rooms later saw us installed in the resort of Saalbach-Hinterglemm, warmly welcomed by some very nice people from Red Bull. They presented us with the list of the men and women, mainly winter-sports champions and Austrian celebrities, who would be attending the event, and whom we were welcome to interview. The list was alphabetical and I quickly scanned down, past a couple of skiers whose names I vaguely recognized from *Ski Sunday*, past Austria's legendary DJ Ötzi, all the way to the bottom. No Vettel. I checked back, not under 'Sebastian', either. For that matter, no 'Horner' or 'Marko' or anyone else from Red Bull Racing. However, there was someone related to F1 whose name I recognized at the bottom of the list: Wolff, Toto, at that time an increasingly active shareholder in the Williams team.

While Alex asked the Red Bull marketeers why they thought we had come all this way if no personalities from their F1 team were present, the crew and I melted away to the restaurant, ordered some Wiener schnitzels, and planned the next day's skiing. It's not like we ever had much of a plan A, but I had thought of a workable plan B. 'Don't worry,' I said, 'we can still make a nice feature out of this, the snowmobiles will look crazy buzzing through these tiny streets, this Toto guy is part owner of the Williams team and they won a Grand Prix this year (courtesy of Pastor Maldonado in

Barcelona). So we'll grab an interview with him, and it'll all be good.' Reassured, Alex told the Red Bull PR people that we wouldn't need DJ Ötzi's contact number after all, and we tucked into dinner. The next day was spent filming pieces to camera on the slopes with a final thought from me in a hot tub (mercifully never aired), before the climax of the snowmobile race and our opportunity to grab Toto, the only F1 interviewee we had. If he was curious as to why we were asking him so many questions about so many teams, effectively mining him for all the F1 content we had hoped we would be getting, he didn't show it. He delivered his interview in the impressive and authoritative style that we've all since become accustomed to. He even gave an early outing to his now famous 'like a bullet' catchphrase.

It was what happened afterwards that stuck in my mind. We had climbed a small hill in order to frame Toto's interview in a pretty way with the village in the distance. Once we had finished recording, rather than trudging back down the hill to a waiting glass of glühwein in the Red Bull Energy Station, Toto lingered to chat as the cameras packed up. We talked some more about Williams and its prospects, with Toto's own protégé Valtteri Bottas due to replace Bruno Senna. He then asked me what I thought about the Mercedes team, where they had gone wrong with Michael Schumacher's return, about Lewis Hamilton joining for 2013, and my thoughts on Ross Brawn.

I told him how motivated I thought Lewis would be, and that he would get on well with his friend from karting days Nico Rosberg, and that I thought Brawn was the perfect team boss to get the best out of both of them. Toto nodded, said his goodbyes, but still didn't leave: he just turned and stood, looking out down to the twinkling lights of Saalbach, deep in thought. 'Odd,' I thought, leaving him to it as I carried one of our tripods on my shoulder down to the hotel. I later found out that Wolff had rolled his snowmobile earlier that day, in practice for the evening's race. He was unhurt, but his team then lost in the final to the managing director of a haulage company. I later rationalized that odd moment by thinking that Toto must have been telling himself to calm it down with all the crashes.

But it wasn't that, and a few weeks later I found out why Toto had been behaving so strangely. He had been asked by the Daimler board and Mercedes F1 non-executive chairman Niki Lauda to assess why the works team's effort wasn't going to plan, and why it had struggled so much since winning the 2009 championship as Brawn GP. Toto had relayed his thoughts, and they had been impressed enough to offer him the job of team principal. However, Toto's way was not clear. Firstly, Mercedes already had a team principal – Ross Brawn. Wolff proposed that he would come in as 'executive director' in one of those opaque managerial structures that F1 teams adopt when they're deeply enmeshed in internal power struggles. Brawn would remain team principal, but as Wolff had also been appointed head of Mercedes-Benz Motorsport (Norbert Haug's old job), Ross would now report to him.

Secondly, Toto knew his limitations. He wasn't a technical genius, and to make the car quicker he needed help from the kind of boffin who could knock up a Large Hadron Collider in your back garden. Such a man was Paddy Lowe, whom Wolff recruited from McLaren for the role of Executive Director (Technical) – another impressive title. The third problem was a bit more personal. It was clear to all parties that if he was to be head of Mercedes, Toto couldn't still be a part-owner of

Williams. So he proposed a deal to the Daimler board. He would agree to sell his stake in Williams and step away from the Grove team completely – but in recompense, Daimler would agree to sell him 30 per cent of the Mercedes team. Lauda was also up for investing if Daimler was selling, so he took a 10 per cent share. It was an incredible bit of negotiation that would, as Mercedes's value rocketed a few years later, make Wolff a billionaire.

Having swum alongside them for so long, at the end of 2013 Ross Brawn became a victim of the piranhas. In technical and sporting matters he was as ruthless as they come, but he was ultimately outmanoeuvred by Toto's business instincts, and he left the team (possibly relieved to go back to fishing in calmer waters). In Brawn's book *Total Competition*, written a few years later with ex-Williams CEO Adam Parr, Ross cited a breakdown in trust between himself, Wolff and Lauda. 'I was beginning to deal with people who I didn't feel I could ultimately trust; people within the team, who had let me down already in terms of their approach,' Brawn wrote. 'In early 2013, I discovered Paddy Lowe had been contracted to join the team and it had been signed off in Stuttgart. When I challenged Toto and Niki, they both blamed each other.' Brawn and Wolff duly went their separate ways only to find themselves working together again – somewhat awkwardly – when Ross became Managing Director (Motorsport) of the F1 management organization in 2017, a position that afforded him some power over Wolff and the other piranhas that must surely have given him a small sense of satisfaction.

F1 team principals are now a different breed compared to the days when they were owners rather than just hired hands. Aside from Horner and Wolff, most of the current crop are ex-engineers who have both leadership skills and an understanding of technical matters. The most successful of this new generation over the last couple of seasons has been Andrea Stella, once Fernando Alonso's race engineer at Ferrari, and now running the race team at McLaren.

Stella doesn't need to swim with the piranhas as he has the McLaren Group CEO Zak Brown to do that. Another former racing driver, Zak took a break from driving to start his own motorsport marketing business and did very well out of it. When McLaren's Bahraini owners needed someone to do the impossible and replace the man who defined McLaren, Ron Dennis, Brown stepped up. He has been very successful, recruiting the right people, getting the money in the door, and reserving just enough time in the day to think up ways to annoy and destabilize his fellow piranhas. The spirit of the club lives on.

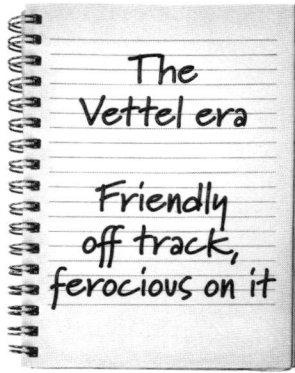

Chapter 16

Sebastian

It's 29 July 2006 at the Spa-Francorchamps circuit. As is so often the case even in the summer months, a passing rain shower has soaked what remains one of the sport's most challenging venues. However, this isn't the Belgian Grand Prix weekend – it's a sparsely attended World Series by Renault event, featuring a grid full of F1 hopefuls.

Among them is a young man who already knows that he's got a better chance of making it to the top than most of his fellow racers. Currently starring in the European F3 Championship, he's been parachuted into the Renault series by Red Bull, his sponsors who are actively nurturing his career, in order to gain some useful extra experience in a faster car. He's already won two races on his debut weekend at Misano in Italy, but Spa is a much tougher track to master.

The rain eases off, but as the race gets underway the track is tricky and wet. The Red Bull car runs slightly wide on the right-hand kerb on the steep slope of Raidillon, just after the compression of Eau Rouge. It spins and clouts the left-hand tyre wall, sending pieces of debris flying across the track. It's a nasty crash, but not life threatening. The driver jumps out and hops over the tyre barrier. He's holding his right hand as a rescue marshal escorts him down the hill to the circuit medical centre. 'This way,' instructs the marshal. 'What's your name?' 'Vettel. Sebastian Vettel.'

The crash was a real blow to Vettel given the career momentum he was enjoying. A piece of carbon fibre shrapnel had flown into the cockpit and sliced through his glove and index finger, taking the top of his fingernail off and leaving a cut that went right down to the bone. The Spa doctors stitched the wound, bandaged the hand, and sent him on his way. He was annoyed at being ruled out of racing for a few weeks, but not as annoyed as Alx Danielsson, a Swedish driver who was lucky not to be killed after hitting an errant wheel from the accident. 'Vettel was driving like it was a rental car,' the Swede told *Autosport* magazine. 'It was just a matter of time before he crashed.' It was a sentiment that would become a common theme in Vettel's early career.

The next day Sebastian drove back to the circuit to catch up with his team. After seeing the large white bandage and splint on his right finger, his mechanics quickly made their own outsize replicas from kitchen-paper rolls and stuck them on their right index fingers in an effort to cheer their driver up. It worked. As soon as the finger healed he was back winning F3 races again, holding up his still-mangled digit on the podium as a joke for his mechanics.

Born to a middle-class family in the small town of Heppenheim in the middle of Germany, Vettel's enthusiasm for motor racing was inspired by his father, Norbert. Much like José Luis Alonso, Fernando's dad, Norbert Vettel was a massive Ayrton Senna fan, and thought he'd give his oldest son a try in karting. Sebastian was a natural – he turned out to be one of those rare geniuses. There wasn't really anything to explain it, he was just incredibly quick, and was soon winning karting trophies. Red Bull were on the alert for promising young drivers, and in 1999, at the age of just 12, Sebastian was signed up for them as a junior driver. Later he would also attract the support of BMW, and thus early on he had two powerful organizations pushing his career – as well as a famous mentor by the name of Michael Schumacher. The two became close after Sebastian won a karting event at Michael's kart track in Kerpen, Germany, and there is a famous photograph of Michael in his prime next to a young Vettel nearly outsized by the two trophies either side of him.

The feeling I always got from Sebastian was that even for an F1 driver, he was a man in a hurry. It was like he'd pressed a fast-forward button on his life. Indeed, by the time of his Spa accident, he'd already tested an F1 car, BMW having arranged for him to drive a Williams at Jerez in September 2005. BMW moved to Sauber for 2006, and, it was clear, wanted Vettel to be part of its long-term plans, even though they already had three drivers on their books.

But Seb had a couple of lucky breaks. BMW had inherited Jacques Villeneuve's contract when it bought Sauber, but by the summer of 2006 the Canadian had fallen out with his new bosses, who had asked him to sit out a race so they could try out test driver Robert Kubica. Seeing the writing on the wall, Villeneuve made his decision and left the team abruptly after the German GP in July. Kubica was duly promoted to a race seat at the next event in Hungary. The Pole, never short on confidence, was asked in that weekend's press conference what he thought Sauber's reasons might have been in choosing him over Villeneuve. 'Maybe the pace,' he replied, to general laughter in the room. His promotion duly opened up a vacancy in the testing role, and BMW was quick to award it to Vettel.

My first encounter with him was at the Turkish Grand Prix, exactly a month after his Spa crash. With characteristic efficiency, the BMW press officers made Sebastian available for interview on Thursday. None of my producers at ITV had heard of him, so there wasn't much demand from the production. However, I was interested in him and the story with the finger, so I thought I'd go along anyway, just to meet him and find out a bit more. When I arrived, Sebastian had just finished with the print media and was ambling down the steps of the Sauber motorhome wearing faded jeans, a white BMW shirt that looked one size too big, with a mop of messy blonde hair underneath what appeared to be a borrowed team cap. I don't remember any other TV crews being present.

He looked like he'd just graduated from high school and was slight, like a stiff breeze would blow him over. My opening question was essentially, how old are you? He'd just turned 19, old by today's F1 rookie standards, but he looked younger. We talked about what his targets were, what he thought was going to be the plan for the weekend, the usual stuff. I asked him about the finger. His eyes brightened in a way I would come to know well over the next 15 years, and he seemed pleased to know that I'd done my research and was aware he'd injured it. He shrugged and said, 'Yeah, it's fine now,' although I could see that the fingernail was still in pretty bad shape.

What he said next really struck me: 'I still don't have much movement, but I came back and won both races at the Nürburgring.

So it can't be that bad.' What I remember is going away from that interview thinking this guy has the kind of inner confidence and clear-headed thinking that you tend to notice in the very best drivers.

A day later Sebastian jumped into the Sauber and notched up the first of many records that he would hold in F1, and one that stands to this day. He became the driver who incurred the quickest penalty in an F1 career when he was clocked for breaking the pit lane speed limit just six seconds after he'd left the garage for the first time, but the rest of the session went well, and overall he made a good impression. At this point, however, there was no race drive available. After being BMW Sauber's Friday driver for the last four races of the season Vettel continued his testing role in 2007, while Red Bull also moved him full-time into the World Series by Renault.

An unexpected chance to step up to a BMW race seat came at that year's US Grand Prix. Kubica had a huge crash at the Canadian GP and was ruled out of the following week's event at Indianapolis. Vettel stepped in, finished an impressive eighth and set another record by becoming the youngest ever points scorer at the time (the record is currently held by Max Verstappen for his six points at the 2015 Malaysian Grand Prix). Kubica was fit and back in the car at the next race, but Vettel's performance had made it clear to long-time backers Red Bull they had a special talent on their hands, someone who should be driving full-time in F1. Red Bull's young driver supremo Helmut Marko rarely needed much encouragement to move his charges around. At the time Californian Scott Speed was having a few difficulties living up to his name and had a major fallout with Franz Tost and Gerhard Berger, bosses of Red Bull's junior team, Toro Rosso. Despite the importance to Red Bull of the

American market, Speed was let go, and Vettel was handed his seat for the Hungarian GP. The move marked the end of his relationship with BMW Sauber – and he never looked back.

He settled in quickly at Toro Rosso, and learned from Tost and Berger, but it wasn't all plain sailing. At the soaking wet Japanese GP, colliding with any car wouldn't have been great, but Vettel made the mistake of taking out Mark Webber in the senior Red Bull while driving under the safety car, putting them both out of the race. The clash with the man who would later be Sebastian's teammate produced a great quote from the Aussie, who shared his exasperation live on ITV: 'Well, it's kids, isn't it?' he said. 'Kids with not enough experience to do a good job, then they fuck it all up for everyone else!' However, Vettel redeemed himself in the team's eyes with a fourth place the following week in China.

There were a few more incidents in early 2008, but he learned as fast as he drove, and soon became a regular points scorer with a decent car that was in effect a clone of Adrian Newey's design for Red Bull Racing. Veteran Toro Rosso engineer Giorgio Ascanelli had worked with Ayrton Senna during his last few seasons at McLaren, and the Italian was key in shaping Vettel into the finished product. Together they found a way of setting up the Toro Rosso which made everything click. After that, he was quick everywhere, and he truly announced his star quality at his team's home race at Monza.

It was a magic combination of factors. Many other drivers struggled in the rain, but that year's Toro Rosso was particularly good in the wet. This, allied with Vettel's pure talent, saw him become the sport's youngest pole sitter in one of the wettest qualifying sessions in memory. Saturday surprises are not uncommon, but what happens in a wet qualifying often doesn't follow through once the weather has cleared up for the race. In this case the rain continued into Sunday, enabling Sebastian to put in a superb performance and win the Italian Grand Prix. It was one of those drives that convinces you, should you have any doubts, that here is somebody who is quite simply at one with an F1 car. As we marvelled at the sure-footed confidence he displayed in the near terrifying conditions, Martin Brundle felt sure he'd witnessed something special. 'This kid is going to be a world champion one day,' he said. It's no coincidence that F1 legends like Ayrton Senna and Michael Schumacher also scored their first victories in rain-affected races.

It was during 2008 that I really got to know Sebastian, usually by messing around with him at our Thursday catch-ups in the paddock. Before the win at Monza put him on everyone's radar there wasn't that much media interest in him, so often it would just be me and one German-speaking TV crew. Once they'd finished, Seb and I were left to exchange stupid jokes or discuss bits of news going on with other teams. He was still young, so I'd sometimes play tricks on him, such as asking his thoughts on FIA president Max Mosley, at that point very much in the news. Seb looked at me uncertainly, 'Erm . . .', then at the camera, then towards Toro Rosso press officer Fabiana Valenti who wasn't helping him out, then back to me again: 'You can't ask me that, you bastard!'

It was no surprise that when David Coulthard retired it was Vettel who replaced him alongside Webber at Red Bull Racing for 2009. Initially the car didn't have the double diffuser that was allowing Button and Barrichello to outpace everyone, but he still managed to score RBR's first win in China, Vettel's second time on top of the podium in F1 – and out came the finger. As the performance advantage of the Brawn waned, Vettel became a regular race winner, and he ended the year as title runner-up to Jenson Button.

The 2010 season featured one of F1's great championship showdowns in Abu Dhabi. After the penultimate race it looked like Fernando Alonso was headed for his third world title. The Spaniard was leading with 246 points, from Webber on 238 and Vettel on 231, while Lewis Hamilton was also still mathematically in contention on 222. Sebastian went into that final weekend very much as the underdog to win the championship. Being 15 points behind, his was a simple task – he had to win the race and not worry too much about what was going on behind him, or any other permutations of who finished where. In the event, he dominated from pole to take victory, while Ferrari famously made a bad strategy call that saw Alonso come out of the pits and get stuck behind the Renault of Vitaly Petrov. The Russian was quick on the straights, keeping Alonso behind him in seventh all the way to the flag, allowing Vettel – who hadn't led the championship all year – to score an emotional first-title success by a margin of four points.

The true greats of F1 all have a ruthless edge. They are mercenary creatures, hungry for success and selfish in pursuit of it. Vettel seemed to be an exception – an easy-going, funny, amiable young man with a social conscience. However, at the Malaysian GP at the start of 2013 we saw a side of Vettel that we hadn't seen before – a hard edge that was reminiscent of his close friend and mentor Michael Schumacher.

The pit stops played out and Mark Webber had just managed to stay in front of Vettel. However, it would be very tight for both men to make it to the end of the race on their remaining sets of tyres, especially Webber's, which had already been lightly used when fitted. To protect both cars and their engines in the Sepang heat, Vettel was told to keep behind Webber for the run to the flag via the radio codeword 'Multi-map 21'. Meanwhile, the same coded message assured Webber that he would not come under threat from his teammate. Teams have radio codewords for all kinds of things – rivals are always listening, so anything that can be done to throw them off the scent is beneficial. To another team, 'Multi-map' could easily have referenced a switch on the steering wheel. It actually referred to the driver order, and '21' meant that car number two (Webber) was being told to finish ahead of car one (Vettel).

However, this was the day that Seb made it clear that he had no interest in team orders. After instigating a spectacular fight that lasted for several corners, he turned his engine up and overtook Webber, going on to win the race. On the podium it was clear that both men were unhappy with the way the race had been managed, with Vettel chafing at being told to stay behind, and Webber annoyed with Vettel for disobeying team orders and overtaking him. Mark then gave an explosive interview to Martin Brundle about how 'In the end, Seb made his own decisions today, and will have protection as usual, and that's the way it goes.'

Vettel looked embarrassed and angry at having been so publicly made to look the bad guy. But, as always in F1, there was a backstory that was just as fascinating. It went back a few months to the championship-deciding Brazilian GP at the end of the previous season, which was the last time Red Bull engineers had used the 'multi' team order codeword. That day, it was Webber who had been told 'Multi-map 12', i.e., that Vettel should be the lead driver and

Webber should let him pass. He initially ignored the instruction, responding on the radio, 'Which switch is that, mate? Which switch, where is "Multi?"', a delaying tactic that went down badly on the Red Bull pit wall.

Codewords were abandoned a few seconds later when Webber's engineer spelled it out on the radio. 'Let Sebastian go, please, Mark.' At the time, Vettel had not been particularly impressed with that, or with Webber's driving earlier during the race, when the Australian came close to putting his title-challenging teammate in the wall off the start line. Looking back on it years later on the F1 *Beyond the Grid* podcast, Christian Horner observed that it looked like Mark was driving for Ferrari that day, and confirmed that Sebastian had been furious with Webber about what happened in Brazil, and that ignoring 'Multi 21' in Malaysia was payback for it. For Vettel, revenge was definitely a dish best served cold.

At the race following 'Multi 21', in China, Red Bull had made Vettel available as usual on Thursday, and I remember walking into the team's hospitality building in the Shanghai paddock with Lee McKenzie. Based on the kind of repair job that the PR people had tried to do the week before, we were anticipating that Sebastian was going to be apologetic, saying something like, 'Well, maybe I made a mistake, and maybe I should have let Mark win, and I won't do it again.' Whether he was going to be apologetic or come out swinging, we knew we didn't want to miss it. Vettel didn't let us down. He was both business-like and totally unapologetic, saying: 'The bottom line is I was racing, I was faster, I passed him, I won.' Silence ensued in the room. I nudged Lee's arm. I'll always remember the unemotional way he said it, and at that moment

I thought: 'Good on you, Sebastian. That's a real racer's hard edge right there.' Vettel was still annoyed with Webber's unguarded remarks on the Malaysian podium, and how that had made him look. He had been back to the Red Bull Racing factory to apologize to the team, but now was clearly not in the mood to make nice.

It's not that Vettel didn't have a soft edge as well. He loved the fact that when he clinched his second world championship in 2011, he did it at Suzuka, where his heroes Senna and Schumacher had won in the past. He appreciated F1's history more than most of his contemporaries, as evidenced by the fact that he could name every world champion from 1950 onwards. He would make a little history himself by going on to win four titles in a row with Red Bull, which was a remarkable achievement, matched only by Juan Manuel Fangio, Michael Schumacher and Lewis Hamilton before, and Max Verstappen since.

Looking back, I think we didn't really appreciate how good Seb was, because he won so much so quickly. When he was in tune with the car, he would reliably qualify on pole, dominate the race, make his way through the pit stops, and close out the win. At the end of 2013 he was still on fast-forward, but the following season would see the pause button pressed. In the V8 engine era Red Bull's partnership with Renault had been very successful. But when Mercedes turned up in 2014 with by far the best power unit for the new hybrid V6 regulations, the Red Bull/Renault package was suddenly no longer capable of challenging for the world championship. Meanwhile, Daniel Ricciardo had replaced Webber and won the three races that were not claimed by Mercedes that year. Seb logged no victories, and had bad luck with mechanical failures.

It was increasingly clear that his four title-winning seasons had taken a mental toll. He started to look around for a career change that would freshen him up. One slight problem was that he still had a year to run on his contract as a Red Bull driver, but crucially it had an exit clause that allowed him to leave at the end of 2014. The way Vettel exercised this over the weekend of the Japanese GP was dramatic.

It was shaping up to be an ordinary race weekend in the Red Bull camp, when completely out of the blue on Friday evening, Christian Horner received a text from Vettel asking him to come to his room at the Suzuka Circuit Hotel. According to Horner, Sebastian was close to tears, explaining how the season had really knocked his confidence and how he wanted to leave the team at the end of the season for a new challenge. It was difficult after all the success they'd enjoyed together, but as Vettel had wisely said on team radio after his third world-championship win in 2012, 'We have to remember these days, because there's no guarantee that they will last forever.'

So with Seb's tears drying on his pillow and after a late night for Red Bull's communications department, Saturday morning at Suzuka saw the news revealed. It was confirmed that Vettel would be leaving Red Bull to join Ferrari as teammate to Kimi Räikkönen, and that Toro Rosso's Daniil Kvyat would be his replacement.

A reboot is always good for a driver who has been with the same team for a few years, and Vettel slipped easily into his new home. He loved Ferrari, and they loved him. He created an F1 meme about how everybody is secretly a Ferrari fan when he joked with Lee McKenzie as she tried to conduct a serious interview with him in Canada. 'Everybody is a Ferrari fan. Even if they're not, they are a

Ferrari fan.' We knew what he meant. Unfortunately, Vettel had the bad luck to be at Ferrari during a period of domination by Lewis Hamilton and Mercedes, and he never did win that fifth title, although it's easy now to forget that he finished runner-up in the championship in both 2017 and 2018, winning five races in each of those seasons. His stay at Maranello ended on a lower note with an uncompetitive car in 2020.

We saw a much more mellow Vettel in his final team, Aston Martin. He enjoyed two solid if unspectacular seasons alongside Lance Stroll. There were some good drives, including a second place in Baku in 2021 and another in Hungary that was then lost to disqualification for a minor weight infringement. What really caught the eye during those seasons was his emergence as an elder statesman of the sport. His work with the Grand Prix Drivers' Association on matters such as safety and governance began during his Ferrari years, and his experience gave him the power to achieve a great deal in those areas. He also became much more vocal about his wider interests in the environment, biodiversity and climate change, sometimes ruffling the feathers of those in authority with his outspoken campaigns on the issues he cared about. He wore a T-shirt at the Miami GP which read: 'Miami 2060 – 1st Grand Prix Underwater – Act Now or Swim Later'.

He was also active in using his platform to campaign for social justice, and was one of the first drivers to unhesitatingly 'take the knee' as a gesture against racism during pre-race ceremonies, while encouraging his fellow drivers to join him. In May 2022 he even appeared on the BBC political debate show *Question Time*, where he acknowledged the hypocrisy of being an F1 driver who effectively burns petrol for money campaigning on the climate crisis. However, he made a point of addressing his personal carbon footprint by offsetting his flights and even – when possible – travelling on public transport.

Vettel always did things his way, and he surprised everyone by creating his first social media account primarily to announce that he would retire from F1 at the end of the 2022 season, at the age of just 35. In each of his final seasons with Aston Martin he finished a modest 12th in the world championship, but his overall career statistics remain impressive – four world championships, 53 GP victories and 57 poles is quite an achievement by any standards. It could be argued that he retired too young, but when you've lived your life on fast-forward, things tend to come around earlier than you expect. Whether it's building bee hotels at Suzuka to promote biodiversity in Japan, or recycling plastic waste in Brazil, Sebastian enjoys being around enthusiastic, interesting people who are passionate about what they do. As far as I was concerned, that feeling was entirely mutual.

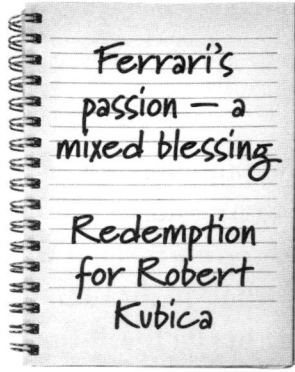

Chapter 17

Nearly Men

Former McLaren boss Ron Dennis always said that in F1 second place was just 'first of the losers'. He was right up to a point, but Dennis himself is proof that, just because you're not winning, doesn't mean that you're a loser. Every driver is competitive, but a common trait I've found in all the elite winners I've reported on – such as multiple world champions Fernando Alonso, Sebastian Vettel and Lewis Hamilton – is that they're so addicted to success and so used to that winning feeling that if they find themselves off the top of the podium for any length of time they are pretty quick to seize an opportunity to move teams, hoping to recapture winning form elsewhere.

All three of those drivers moved to the Scuderia Ferrari in search of their next world championship. On paper the iconic racing team promised Alonso and Vettel so much, but neither managed to win the title. In 2025 Hamilton followed in their footsteps, and at time of writing, alongside teammate Charles Leclerc, is finding it hard work to end a drivers' title drought that extends back to 2007.

Alonso had a habit of moving teams as he tried to position himself between cars that could win and team bosses he hadn't fallen out with. A victory on his Ferrari debut at the 2010 Bahrain GP set the scene for what ought to have been a third world championship that year, but for the ill-timed pit stop and Petrov's slippery Renault at the last race in Abu Dhabi that handed the title to Red Bull's Vettel. Ferrari then endured a shockingly disappointing 2011 season after a correlation problem between the Maranello wind tunnel and the track resulted in a car that was very difficult to drive. For the first half of the year Ferrari churned out new parts, but they didn't work. As Alonso said, 'We found ourselves in April with a car that was slower than what we had in February testing. We understood this too late, and that cost us the championship.'

Things improved in 2012. Alonso won three races and finished runner-up to Vettel by three points at the end of a season that saw seven different drivers win the first seven races. Fernando and his engineer Andrea Stella tried to make it work at Ferrari, but in subsequent years the team started to change. Stefano Domenicali took responsibility for the disappointing performance of the hybrid engine in 2014 and left the team. It was an honourable thing for him to have done, given that there had been pressure from the Ferrari president Luca di Montezemolo to sack engine chief Luca Marmorini for the power unit's deficiencies, something Domenicali refused to do. Marmorini left the team that summer anyway, to be replaced by Mattia Binotto – a name we'd hear more of further down the line. But when Di Montezemolo replaced Domenicali with a complete unknown, it signalled the beginning of the end for Alonso.

Marco Mattiacci had been head of Ferrari's operations in North America, and as such was essentially a successful road-car salesman. Given America's fondness for Ferraris and the fact that the company limited supply in order to maintain the brand's exclusivity, it was probably hard to underperform in such a role. However, after losing Domenicali, Di Montezemolo needed a new team boss in a hurry, and didn't think that Mattiacci's lack of technical knowledge or F1 experience would be a problem.

On a cloudy Shanghai Friday in April 2014 Mattiacci turned up for his first day at work. By any objective measure, he cut a stylish figure in the paddock, carrying a smart tan leather bag and wearing a pair of trendy yet practical sunglasses, the kind that fold up at the bridge for easy storage. Not that Mattiacci showed any sign of folding them up.

'Hey, the new Ferrari boss looks quite cool,' I remember remarking in our morning production meeting. The first practice session started, and the new boss watched from the garage. It got a bit darker as the day wore on, and Shanghai's famous smog hung in the air. The sunglasses stayed on. Ferrari's efficient media office invited us to a press conference with the new team principal. Mattiacci entered the room. We all looked around and smiled. 'We're indoors. Is he going to take the sunglasses off? No? Really?' It became quite the joke around the paddock, with some calling him 'Hollywood'. I preferred to imagine him taking cues from Enzo Ferrari himself, a man famously keen on dark glasses. Mattiacci was later asked directly why he had been wearing sunglasses all day, even indoors. Unfazed, the genial Italian replied: 'It's a very good question. In particular, if you do in less than four days almost 40 hours of flights, and you don't sleep, probably you need sunglasses!'

It was a decent enough explanation, and Mattiacci turned out to be a perfectly capable manager. However, things didn't end well for him at Ferrari either. He fell out with Alonso, and signed Vettel from Red Bull as his replacement. Vettel, then in his fallow period, with Hamilton and Mercedes winning consistently, had decided to try to succeed at Ferrari where Alonso had failed. Mattiacci hinted at a lack of energy from Alonso when he said of Vettel, 'He will bring the enthusiasm needed to go through certain difficult moments.' That prompted a terse response from Fernando, who noted that Mattiacci had 'only been here for a few months and has not seen the five years that I've spent here and how I've fought every single race.' Alonso had a point. After Monza, Di Montezemolo resigned as Ferrari president following disagreements with parent company FIAT's CEO, Sergio Marchionne. With his mentor Di Montezemolo gone, Mattiacci was also ousted as Ferrari team principal after just eight months, to be replaced by Maurizio Arrivabene, who had worked for many years with Ferrari sponsor Marlboro.

FIAT boss Marchionne was already a very powerful man in the motor industry, and after Di Montezemolo's departure he started to take more of a personal interest in F1. Everyone at Ferrari was terrified of him, and his occasional visits to the paddock would noticeably send the whole team into a mild panic. As I wasn't directly invested in the fortunes of Ferrari, to me he appeared more like a kindly uncle in his round glasses and trademark casual wool jumper, and I quite looked forward to his appearances. I have no idea how, but we struck up an unlikely professional relationship. It was probably because he wasn't my boss, and we never really needed an interview with him, so the stakes were low as to whether he told me to get lost or not.

This is what used to happen. We'd get word from our Italian colleagues that Marchionne was going to pay a visit to the paddock for meetings and to inspect the Ferrari troops. My producers would pass this on with a 'we don't really need him, but keep across it' instruction. Anticipating a bit of an Italian media scrum, I'd wander down to the paddock with my cameraman Lee. You might well have heard me or Martin or Natalie talking about Lee, or 'The Marine'. Not difficult to guess from the nickname, but he served in one of the most feared fighting forces in the world, the Royal Marines, and saw action all over the globe. I'd trust him with my life.

An interview with Sergio Marchionne was never going to be too dangerous, but as a former Marine, Lee likes getting stuck in. So we always looked forward to being involved in the drama whenever Marchionne was around. We'd get ourselves in among the crowd outside the Ferrari motorhome, and when the boss came out, there would be a lot of jostling from his bodyguards. We'd allow ourselves to be pushed around by these guys in sunglasses a little bit before I would pipe up loudly, in as British an accent as I could muster, 'Ah, good afternoon, Mr Marchionne.' He would look up and say, 'Good afternoon,' and deign to give me a brief interview.

After the second or third of these encounters, he'd come to recognize me and Lee, and began to single us out and greet us with, 'Ah, hello, how have you been?' or some such pleasantry. Sometimes what he had to say was newsworthy, and we'd use a soundbite in our programme. When Arrivabene was appointed as Mattiacci's successor he would always be pinned to Marchionne's side during our interviews, and would literally stand there growling at me as I exchanged pleasantries with the man who could fire him at a moment's notice. Ferrari's corporate press liaison Stefano Lai (pronounced 'lie', a somewhat unfortunate name for a press officer) always looked down on proceedings with a little smile, although being 6ft 5in, he looked down on most people. Marchionne died in 2018 from complications following shoulder surgery. He had his enemies, but I was saddened by his untimely passing.

By this point Fernando Alonso had left Ferrari. He had become convinced that the Scuderia could not, in the short-to-medium term, win the world championship, and that they did not have the leadership or the technical expertise with which to beat Mercedes. There didn't appear to be a seat free for him at Mercedes, even as a replacement for the retiring Nico Rosberg. It was felt that the pairing of Alonso and Hamilton would have created too much intra-team tension – a real pity for the rest of us who wouldn't get to see their 2007 McLaren rivalry re-kindled – and there wasn't a seat at Red Bull, who were focusing on Max Verstappen. That left Fernando with the unlikely option of a return to McLaren – quite astonishing given the circumstances under which he'd left eight years earlier. But by then Martin Whitmarsh had departed, replaced by Eric Boullier, who had done a decent job at the Enstone-based Lotus team, and Ron Dennis would lose a boardroom battle and effectively be forced out soon after. Besides, their new Honda engine couldn't be that bad, could it?

Sadly, it could. Honda rushed into their F1 return before their technology was really ready, and its engineers were forced to squeeze their power unit into McLaren's 'size zero' rear end to please the aerodynamics department. Achieving these constraints compromised the power unit's architecture, rendering it chronically unreliable. It was beset with high internal friction and vibration problems, plus the hybrid system kept failing. Engineers tried their best to fix the issues, but with little success. What made things worse was that Honda set ambitious targets, but then repeatedly failed to meet them, which eroded McLaren's faith in their technology. And so, after three years, tacitly admitting that the 2015, 2016 and 2017 campaigns with Honda power had been a waste of time and money, McLaren decided to end the partnership.

The story of how they did it, especially considering Honda didn't want to leave, was F1 politics at its complicated best. Essentially it involved McLaren giving Red Bull the Honda engine (something that would work out very well), Red Bull giving Carlos Sainz to Renault (replacing poor Jolyon Palmer before Sainz himself left a year later to join McLaren) and McLaren finally getting a supply of Renault engines.

It had been only too easy for Eric Boullier to blame Honda, but the McLaren chassis had also fallen behind and the Frenchman left the team in the middle of 2018. Where was Alonso in all this? Fed up and disillusioned. He left F1 at the end of 2018, aiming to become the most versatile racing driver of the modern era. His talent, wasted over those three years at McLaren, turned to the Dakar Rally, the Indianapolis 500 and helped Toyota to win the Le Mans 24 Hours and the World Endurance Championship. His roving quest to become the 'most complete driver in the world' was entirely validated as he found job satisfaction in other categories of motorsport.

I thought he was done with F1 for good and even made a 'goodbye letter to Fernando' feature that we filmed in producer Jack McShane's flat. What I had underestimated was just how much Alonso still wanted Grand Prix victory number 33, and world championship number three. He returned to his roots at Renault (then under the Alpine brand) in 2021 for what was his third stint with the Enstone outfit. It wasn't a success, and in the middle of the

2022 season he stunned the F1 world by signing for Aston Martin as replacement for the retiring Sebastian Vettel. As I write, he's still enjoying an astonishing career, and it's not over yet.

Alonso may have followed Vettel to Aston Martin, but at Ferrari it was the other way around. The German arrived at Maranello in 2015 and duly won his second race with the team. Unlike Alonso, Vettel did believe Ferrari could challenge for a world championship, but his 2017 campaign effectively ended on the damp streets of Singapore. He was on pole, but after a brief rain shower on the grid he drifted left off the start line and crashed into Max Verstappen and his own teammate Kimi Räikkönen. A week later in Malaysia an inlet manifold failed in qualifying, leaving him fourth in the race. Another retirement followed when a spark plug failed in Japan.

With at least 50 points dropped over three races, defeat was inevitable, but that didn't make it any easier to take. Ferrari people were in tears when Hamilton wrapped up the world championship in Mexico. Passion. It is Ferrari's strength, and its weakness.

Vettel was pretty down, too. He knew the unreliability wasn't his fault, but the unforced errors – a road-rage sideswipe on Hamilton in Baku that earned him a penalty, and the start line crash in Singapore – were. I asked him in an interview if he felt frustrated about those two lost opportunities. He made the distinction between frustration and disappointment. In Seb's mind, frustration would have been like some anxiety dream where he was in a position to win, but something was holding him back. Disappointment was when things simply went wrong, like crashing at the start or the car blowing up. So, he said he was disappointed, but not frustrated.

'I think I see what you mean,' I said, when the cameras had stopped rolling, and our sound man Dave Haigh had pulled the microphones from the inside of our shirts.

'Yeah, and it's fine,' replied Vettel. 'I'm a big believer that things happen for a reason.' Forgetting for a second that we weren't two friends down the pub, I replied, 'Sure, Seb, but you know the saying about everything happening for a reason? Sometimes that reason is that you're stupid and make bad decisions.' Vettel raised an eyebrow, turned and wandered off to his next media engagement. I wondered if I'd perhaps given him something to think about, although most probably he just thought I was a prat.

Vettel left Ferrari at the end of 2020 with an impressive tally of 14 wins, but no world title. There were many reasons why it didn't work out, but I felt one was that he missed the firm guiding hand of Christian Horner, and the even firmer hand of Helmut Marko. At Ferrari the drivers are the stars, and at the time, the bosses kept them on a much looser rein than Seb was used to at Red Bull. There was a lack of consistent leadership. Marco Mattiacci had signed Vettel, but he was moved aside shortly after, and his successor Maurizio Arrivabene had enough on his plate finding his own way, so he tended to leave the drivers to it. Mistakes started to creep into Seb's game, costing him points over the 2017 and 2018 seasons, and he started showing a level of impetuousness that hadn't been in evidence (and, knowing Horner and Marko, would have been cracked down on and prevented) at Red Bull. It is often evident in F1 teams that drivers' behaviour is affected by the example of the team boss. If drivers have a strong boss that they respect, and are a bit scared of, they stay in line. If they have a weak boss who they're not afraid of then they're much more likely to just do as they please.

One driver who could surely have been a world champion for Ferrari in the 2010s was Robert Kubica. In many ways, Kubica is F1's greatest lost talent. He was quite simply incredibly quick. In his teens he moved from Poland to further his karting career in Italy. He learned the language, and the country became his second home. His speed meant that he was always going to find a place in F1, and his chance came with BMW Sauber when he replaced Jacques Villeneuve for the last third of the 2006 season.

The following year in Canada he suffered a setback when a small mistake – touching his front left tyre with Jarno Trulli's right rear – had a huge result. A high-speed smash into a concrete wall and a barrel roll down the track. I watched the accident from the McLaren garage, and every single person in there, hardened mechanics and engineers who'd seen it all in F1, could barely look at the sickening replays of the crash – the violence of the impact and the car's destruction was scary, especially as you knew that Kubica was right in the middle of it.

Exactly 12 months later at the same venue he could laugh about it – not only did he recover, but he returned to the track to win his first Grand Prix and took the lead of the 2008 drivers' world championship by four points. He believed that he had a shot at winning the title, but his BMW bosses halted development of the car in favour of the following year's contender, and the team's form faded. At the end of 2009, however, BMW pulled out of F1 and Kubica moved instead to Renault for 2010, picking up three podium finishes.

Ferrari appreciated how good he was and, encouraged by Fernando Alonso, offered Kubica a kind of provisional race contract to give the Maranello squad first call on his services for the 2012 season as a potential replacement for Felipe Massa. All he had to do was see out one more year in a midfield Renault and he'd likely move to Ferrari. However, his life and his career path were to change in February 2011, just a few weeks before the start of the F1 season.

While his peers were sunning themselves on beaches or swooshing down alpine ski slopes, Robert was keeping his driving skills sharp. He entered a small Italian rally, the Ronde di Andora, on the country's north-western coast, only an hour's drive from Monaco. He was just as quick in a rally car as he was in F1 machinery. But on the very first stage, his Skoda went wide on one particular corner and hit a metal Armco crash barrier which punched a hole in the engine bay, pierced the cockpit and ripped into Robert's right side.

His co-driver was unhurt, but the injuries that Kubica sustained to his leg, arm and hand were so severe that his insurance company determined they would be career-ending, and it paid out on his loss of future earnings as an F1 driver. Kubica had other ideas. Through sheer bloody-minded determination, numerous medical procedures and operations, while suffering near-constant physical pain and requiring copious amounts of emotional grit, he focused on his recovery. He returned to rallying in 2013 and then moved to GT racing. Rally and sportscars had room in the cockpit for him to extend his elbow enough to have a good amount of movement in his right forearm and hand, but a single-seater was always going to be another matter.

He'd come this far, so nothing was going to put him off. Six years after the rally accident, Robert Kubica got a second chance at Formula 1. It was just a private test in a five-year-old car, but his training paid off, and he was able to drive a 2012 Renault reasonably quickly and without fundamental limitations. He could only change gear via a single paddle pushed or pulled by his left hand, and he lacked the fine-motor function to change steering-wheel dials with his right hand, so he also did that with his left. He did enough during initial running with the old car in Valencia and Paul Ricard to be given a proper test at the Hungaroring after the 2017 Hungarian GP. He was fourth fastest of those running, and completed an impressive 142 laps of Budapest's demanding, twisty track. He adopted a technique whereby most of the driving was done with his left hand, but when he needed to take that hand off the steering wheel to press a button or change a switch, he was able to drive by pushing his right palm on to the steering wheel and moving it with his shoulder muscles.

Renault made it clear that he would only be offered a race drive if he was good enough to win – anything else would be detrimental to both parties. It was close, but in the end, Renault passed on him. By then, though, Williams was interested, and after a good showing in end-of-season testing it looked like his shot at regaining F1 driver status was on, helped by some Polish sponsorship. I caught up with him in December at the Autosport Awards. At 32 he looked in great shape – a few grey hairs on his temples, but the sparkle in the eyes was still there. On stage he talked of how he was in better condition physically than in 2010. 'I have to work much harder now – I was a lazy guy in the past,' he joked, adding that 90 per cent of his driving ability was just as it was pre-accident.

In January 2018 he was confirmed at Williams in a reserve driver role, and that year he made his welcome return to Grand Prix weekends with three Friday practice outings. He did enough to land a Williams race drive for 2019. Eight years after the rally crash that nearly killed him, Kubica was finally back in F1.

He scored a symbolic point with a 10th place in Germany, but he was overshadowed somewhat by young teammate George Russell. He was then dropped by Williams at the end of that season and went instead to Alfa Romeo as a reserve driver, landing two further race outings in his second year there in 2021 when Kimi Räikkönen was ill with Covid-19. But it would be in 2025, 13 years after he should have made his Ferrari F1 race debut, that Robert's dream of winning for the Prancing Horse finally came true. Having moved to the World Endurance Championship, he and his teammates Yifei Ye and Phil Hanson won the Le Mans 24 hours, a remarkable achievement by an extraordinary racing driver.

Kubica was and remains hugely likeable. It's not often that someone gets a shot at redemption after fate has kicked them to the floor. His return was a great story not only because it was unprecedented in F1, but because it granted him a second chance at his F1 dream, and it went some way to putting right the tragedy of a career cut short.

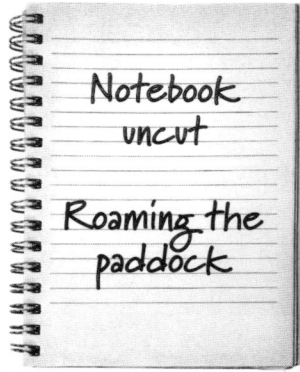

Notebook
uncut

Roaming the
paddock

Chapter 18

Ted's Notebook

It's 6.35pm on a Saturday in May 2022, and I'm standing at the top of an escalator on the mezzanine level of the Hard Rock Stadium, home to the Miami Dolphins N F L team. To my left, cameraman Pete Velluet starts a slow pan across the famed playing field, from scoreboard to touchdown zone, as I enthuse about how well the new Miami Grand Prix venue has accommodated the Formula 1 circus for the weekend. The red light on top of Pete's camera confirms that we are live. The director is 'on' our pictures, and there are no replays or other shots going to air. I can hear myself in my headphones, which means that every word I say is, right at that moment, being heard by millions of people across not only the U K and Ireland but also Australia, New Zealand, South Africa, Canada and anyone tuning in here in the USA on the ESPN network. I try not to think about it.

I'm just finishing my explanation of what a success the first qualifying session has been and am starting to describe how Ferrari have locked out the front row of the grid when I realize that I am, in fact, stuck on this top level of the stadium. I need to be doing the next 22 minutes of the programme on ground level, where all the F1 team garages are, but I appear to have no way of getting back down there. There's a spiralling concrete ramp around the corner, but I'm already at the limit of our radio-frequency reception zone, and even if the picture did hold up, it would take a good three minutes to circle down the walkway to the bottom. Even if I ran down, we'd lose the sound and probably the pictures and effectively fall off-air. Not an option.

While half of my brain ponders this problem, the other half tells my mouth what to say about how this grid line-up should make for a fascinating race. Right. I've got to make a decision. We're live on-air – I don't have a choice. The camera and mic worked when we came up the escalator, so I know they will work when we go down. So what if it's still ascending? We'll just have to out-run it. I check that Pete is up for it. He shrugs his shoulders and nods. Good lad. He's always up for it – he's the kind of person I'd have gladly served alongside in any major conflict of the 20th century, and he wasn't an actual Royal Marine like my other most regular cameraman, Lee. We walk on to the landing zone and go for it. I had noticed on the way up that it was quite a fast escalator, so what worked in our favour around four minutes ago is now, annoyingly, demanding some rapid footwork. I decide to lead the way boldly, figuring that if Pete trips and falls, I'll be able to catch him, or at least catch his camera, which, he would agree, is the most important bit. I do check he's holding the handrail as he runs down after me. Safety first!

We're making good progress, so I launch into an explanation of how Ferrari managed to set their cars up perfectly to be quick on

Miami's long straight, yet still have the downforce to make up time in the corners. A blazing lap by Charles Leclerc put him nearly two-tenths of a second ahead of his teammate Carlos Sainz, and rival Max Verstappen. Three quarters of the way down now, and amazingly I'm not even out of breath – all that gym training is paying off! In my headphones I can just about hear some voices in the control room, possibly the words 'What is he doing?', but I can't say for certain. By that point, anyway, we've reached the bottom, jumped off the ramp and we're on our merry way down the paddock. We're still on-air, the radio camera picture was solid, no breakup on my microphone, and 21 minutes left on *Ted's Qualifying Notebook*. Sorted!

It's the only part of the weekend's coverage that I am fully responsible for, but weirdly enough, the *Notebook* wasn't actually my idea. It began life as a column for the ITV F1 website that its then editor Simon Strang asked me to write after every race. Simon had the notion that there must be little nuggets of information that I'd heard or observations I'd made that I'd scribbled down in my notebook but hadn't had time to use on TV. As it was a shame to waste them, couldn't I combine them into a column to help keep the numbers clocking up midweek? I agreed and he suggested that we call it *Ted's Notebook*, which seemed fitting.

The website column carried on for a year or so to no great acclaim. Some people must have been reading it, though, because when I moved to work for BBC Sport in 2009, assistant F1 editor Andrew Benson (who was primarily responsible for the BBC F1 website) was keen for me to carry on with it. However, he wanted two changes. Firstly, he didn't want to use the title *Ted's Notebook*, because ITV had done that, and it should be different if it was on the BBC. Secondly, after a year or so as a written column, the time needed to navigate an increasingly crowded race calendar was starting to affect my writing schedule. So together with programme editor Mark Wilkin, Andrew suggested that we turn the column into a video blog that we would record after qualifying or the race and publish on the BBC F1 website. 'Fine by me,' I said. 'But what shall we call it?' Benson said that since it was coming from the pit lane we should just call it *From the Pit Lane*. It ran for about 10 minutes, with me talking to camera, roaming around the pits and paddock, encountering some drivers, and relaying news and information from each team that hadn't made it into the main show.

I didn't come up with the name, and I didn't come up with the format. However, I did enjoy doing it. I felt excited yet comfortable, and I liked the element of surprise in that the viewers didn't know what was going to happen next, because neither did I. So it came to pass that when I joined the team at Sky Sports F1 in 2012, somewhere on our executive producer Martin Turner's 'to do' list was the subject of what he called my 'paddock ramble'. About two months into our pre-production phase Martin said, 'I liked what you were doing on the BBC website, we would like to do something similar – but there's no point calling it *From the Pit Lane with Ted Kravitz*, it's too long and cumbersome. Can you think of anything else we can call it?' I thought for a few seconds as if summoning up some creative spark. 'How about *Ted's Notebook?*'

Martin followed F1, but he clearly hadn't come across any of my earlier columns on the ITV Sport website, so he looked at me wide-eyed and said, '*Ted's Notebook*. Great name!', confirming in that moment that there are no new ideas in television. Turner had worked hard to convince Sky and Formula 1 that rather than just producing coverage when practice, qualifying or the race was on (as had been done on the BBC and ITV), Sky Sports would create a dedicated Formula 1 channel, and he saw an opportunity with the *Notebook* to convey a sense of the last word on the day's events, as well as something that could be repeated on the channel after race replays throughout the week. He wanted the segment to be its own programme, with titles and a set duration. His view was also that you shouldn't have to be a Sky subscriber to watch the *Notebook*, it would also serve as an online promotional tool for the main programme – something to draw people into what had been happening, and which would hopefully make them interested in following the season with us on Sky Sports.

Following a modest start in 2012 the *Notebook* soon expanded to appear after qualifying as well as after the race, and eventually it grew to fit a commercial half-hour time slot – two parts of 13 minutes – with a total running time of around 26 minutes, in order to fit a four-minute advert break in the middle. It's been in that format ever since – as of 2025 we're into our 14th year.

The *Notebook* aims to do, or has evolved to do, two things. Firstly, (and I try to prompt myself to remind viewers of this at the start of at least one show per weekend), it's the programme that tells you what happened at the Grand Prix if you couldn't watch the Grand Prix. If you've got kids, or you've got things to do on a Saturday, or you've got the family round for a roast lunch on Sunday, it might not be possible to slip away to watch qualifying or a two-hour race with an hour's build-up and an hour of analysis either side. I wanted to make something that was easily accessible, a half-hour show that would sum up all the stories and give viewers a feel for what was going on in the paddock, and for it to be available after qualifying or the race had finished.

Secondly, it tells the stories of all 20 drivers. It's simply not possible for the TV director to follow every driver and the stories of their efforts in qualifying or the race. Directors have to decide if there's a frontrunning driver in danger of being knocked out in a qualifying session, or what the closest battle in a race is – these sorts of things tend to be the focus. However, every driver's story is interesting – even if they're having a lonely race. There's a story in every car, and the *Notebook* aims to tell it.

The *Notebook* also sheds light on the unexpected things that happen over a race weekend. Perhaps a pit stop went wrong, a driver's radio failed, the chequered flag was waved too soon, or something notable happened on the grid. I don't want anyone to go away at the end of the day wondering, 'What happened there?'

The *Qualifying Notebook* on Saturday is a bit more relaxed. I'll start with a general look around, giving a sense of the atmosphere at the track and in the host city, what the fan experience has been like, and what the main story of the weekend is. Then I get into the qualifying results, and while I'm reviewing those team by team, I keep an eye out for who I can find, or what I can show around the place.

In terms of notes, I don't use a tablet or phone, but an actual physical notebook in which I jot everything down by hand. What's not well known is that I don't even buy my own notebooks. Pirelli happens to supply the perfect pocket-sized, A6, ring-bound notebook, and they are available at every race in the tyre company's motorhome for anybody in the paddock, so I just pick up a new one when I've used all the pages. I'll use around five notebooks per season. One usually lasts for four or five weekends, but if there's a wet race and it gets soaked, I might have to change early! I'm often asked what I do with my old notebooks after they're full. If there's nothing in them that I might want to refer back to, I might throw them into the recycling bin when I get home. I do keep most of them, though, in a box file, waiting for me to rediscover them at some stage in the future and then throw them into the recycling bin after all.

If it's been an eventful race, the pages will be packed full of scribbles and notes, some bigger than others according to importance. I divide it up with five teams to a page and each driver in a sub-section of each team. There will be a couple of subjects at the top of each page to remind me of things that have happened during the weekend that I might want to start with, but otherwise the pages will pretty much be blank before the action starts. Throughout qualifying I'll fill in the results and events of Q1, Q2 and Q3 as they happen, and then the same for the race. If somebody drops out, I'll write DNF next to their name, put the reason for retirement if we know it, and then just write the odd note about what was happening before they stopped. By the time the qualifying session or race is done, my two pages are half full. I'll list where the drivers finished, and any other details about whether the result changes their position in the world championship. There'll also be a little arrow up or down next to the team name indicating whether the result has moved them up or down in the constructors' championship. I'll also remind myself to put dates and times of upcoming races that I might have to promote at the end of the programme.

The main way I get the information that fills the *Notebook* is by talking to people. The research starts with the usual team boss or chief engineer interviews on the pit wall after the race. We'll usually do three or four of them for the post-race TV analysis, and I'll check in with others if there's something that happened in their races that should be explained. By then the drivers are in the interview pen, and we've already got our reporters there – Rachel Brookes, Craig Slater or Natalie Pinkham. In some ways it would be much simpler if I could also be there to listen to all the drivers and write down their quotes, but as most of the driver interviews are played out on our post-race show before the *Notebook* airs, there's no point just repeating that, unless they've said something particularly important.

My focus, then, is on researching extra background information from the teams – and it does happen that sometimes we're informed of things that the drivers themselves didn't know about during the race. Not everything goes out on the team radio. Sometimes if a driver doesn't have a good race you'll only find out later that they had some kind of technical issue, like a broken floor or an engine problem. Information like that obviously changes the whole story of their Grand Prix; I think it's important to get that information out to viewers so they can put the performance of that driver into context – otherwise they might be thinking, 'He only finished 18th, he was really slow today.'

This is where your contacts come in useful. When the pit lane clears, I'll go to each hospitality unit in the paddock to find anyone who can fill me in on any of the details. Usually, I only need a quick question answered. My first point of contact is usually the media relations or communications officers. I have a very good relationship with Stuart Morrison, the head of communications at Haas – he understands what I'm after. When his team wasn't scoring many points in 2023, I'd ask about their race and he'd simply say, 'No pace, slow, that's it.' The whole exchange would take five seconds, and I had all I needed to know.

There is a provisional on-air start time for *Ted's Notebook*. We seldom hit it, as things tend to over-run after eventful races, but it doesn't matter. When we do get underway, what I do is pretty much unplanned, and that is by design. There have been various attempts over the years to make it into more of a polished TV programme, but the point of the *Notebook* is that once we start, it is live, it's one shot and we don't stop, even if I make a mistake or something goes wrong. We've even had a camera fail in the middle of filming and we've had to pick up with another camera. The shakier and less polished it is, the more I like it!

In deciding where to start I think carefully about how the previous programme ended. I don't want to pick up directly from what Simon Lazenby and the pundits were discussing in the post-race analysis, because I don't want people to start with something they've just seen. To my mind it's good to start somewhere fresh, and I'll have a quick think about what my top story is. It might not always be the race winner; if a team is doing a celebration photo with all the crew in the pit lane, I might start there – my choices are very much based on what's happening in the moment. Team photos are always a good watch, as every team member congregates in front of the garage around their drivers and there are always some amusing moments to witness.

I will always do a few minutes on the event and a few minutes on the winner, and after that I'm only left with around 30 or 40 seconds per driver, which isn't very long to explain what happened in their races, but I make sure I get all the best information in.

Cameramen Pete and Lee know what I'm after in terms of what to film during the *Notebook*. They tend to roam off by themselves while I'm talking, but I'll always have a mental picture of where they are, so I know where to find them if I need them to show something or someone. Sometimes you'll see Pete walking up the side of a team truck, and because I saw him go down there, I'm usually waiting for him when he walks back up the other side. My cameramen are like well-trained puppies in that regard, or perhaps I'm the puppy and they have me well trained!

Because we're just two people, we can cover more ground than a usual programme presentation crew which allows us to nab key people for a quick interview. The sheer joy and relief in Lando Norris when we spoke after his 2025 Monaco Grand Prix win made for a wonderfully human moment, while a swift encounter with Adrian Newey following the 2024 Japanese Grand Prix turned out to mark a turning point in his Formula 1 career.

Red Bull Racing had just dominated the race in Suzuka, finishing first and second with Max Verstappen and Sergio Pérez. The team were holding their victory photo on the grid, but as the champagne started to flow, I spotted Adrian and his wife Amanda slinking off the side of the grid into the pit lane. Dodging a forklift truck to get to them, I caught a very reflective Newey, crediting his team of engineers back in Milton Keynes. We'd later discover that it was on that weekend, maybe on that exact day, that Newey had decided to leave Red Bull Racing, hand the baton on to the team beneath him and head for pastures new.

In Mexico in 2022 I was talking about Alex Albon's race when the camera caught James Vowles, who was then still the chief strategist at Mercedes, chatting with the owners of Williams.

The conversation could have been about anything, but a couple of months later James was confirmed as the new Williams team principal, and I'm convinced that by chance we captured a crunch meeting.

When the opportunity arises, I stop and talk to drivers or team bosses, but those encounters are never planned. By the time we get on-air with the *Notebook*, a lot of the drivers are just about finishing their briefings and are heading back to their hotels or to the airport, and it's fair game to grab them on the way to the car park, if they are willing to talk. Nico Hulkenberg likes a chat, as does Lando Norris, whereas Alex Albon is normally very good at telling me to go and bother someone else! Going further back, Daniel Ricciardo was always prime *Notebook* gold, a flash of eye contact and that grin would play on his lips. He knew that there was mischief to be had on live, unedited TV.

Often while I'm doing the *Notebook* the team bosses are conducting press briefings in the motorhomes for the written media. We don't join those. It's not that we've been told not to, it's just not good form. It's a professional courtesy among the media that after TV has had its time with the team bosses, then it's the turn of the print and website journalists. However, if there's a massive story breaking, such as a big controversy, I've been known to nip into a motorhome and gatecrash a media session.

Sometimes after a race there's a big ongoing story such as a technical infringement and an FIA investigation playing out while I'm on-air. George Russell's eventual disqualification after he had won the 2024 Belgian GP on the road is a recent example. It wasn't exactly investigative reporting on my part, given that the FIA had already issued a document from technical delegate Jo Bauer saying that there was a discrepancy in the minimum weight for car number 63. We happened to go to parc fermé at just the right time and saw the FIA officials wheeling the Mercedes on and off the scales, and the mechanics taking wheels and front wings on and off. Bauer and his colleagues were still in the process of confirming that Russell's car was indeed underweight, and we were there to bring the viewer into the moment, watching it play out in real time.

Post-race FIA investigations like these often take a while, and most don't get concluded before the *Notebook* finishes. With the Russell situation at Spa, for example, I couldn't say that he was definitely going to be excluded. For all we knew there was a valid explanation for the weight discrepancy, or Mercedes might have been able to show that a part fell off or something. So you can never call a stewards' decision in advance one way or the other, because there's always a chance that there won't be a penalty. If the results are still in question, I'll finish the programme encouraging viewers to check in with *Sky Sports News*, and we'll post the result on social media. That can be a bit frustrating, but there's only so long that we can stay on-air for, and the FIA don't consider TV schedules when they're conducting their investigations.

Because my cameraman is often some distance away from me, looking into motorhomes or garages, sometimes other people see me talking, don't realize that I'm actually live, and come up for a chat. The person who did this most often was Britta Roeske, Sebastian Vettel's media and general manager. She had a tendency not to notice that I was live and would say, 'Ah Ted, there you are. I have the details for next week's interview.' 'Thank you very much, Britta, but I'm in the middle of something.' 'Oh, you should have told me!'

Hopefully by now, I think most paddock regulars know what I'm doing when they see me wandering around with a microphone, seemingly talking to myself. Sometimes I notice people in my peripheral vision who look as if they might want to come and start a conversation. If I don't know them, or I can't predict what they're going to say, I'll turn my back and move away, hoping they'll get the message.

We can't edit a live *Notebook*. If you do see an edit or a cutaway on the Sky website or YouTube versions, it's probably because somebody has come and interrupted us or said or done something that we didn't want to include. A team boss's brother once leaned into my microphone and said something about a rival team that was potentially slanderous. It went out live, but we were able to edit it out of the replay. We also had to edit another one where a teenage fan in the paddock came and flicked the v's into the camera!

The regular race weekend *Notebook* has led to a couple of spin-offs. The first is the *Testing Notebook*, where we do a roundup of what happened at the pre-season test at the end of each day. Just as it's difficult for people to watch four hours of Grand Prix coverage on a Sunday, it's even harder for them to watch nine hours of testing coverage on a weekday, so I hope it serves our viewers to have a half-hour roundup of everything that happened. Then we have *Development Corner*, which started at a Barcelona test a few years ago. The track's media centre used to have little soundproofed booths where, before email existed, journalists would file their copy to their newspapers or magazines by telephone. Nobody used them anymore, but they remained in place, so I would pop into one with my laptop, the cameraman would join me, and I'd go through spy photos and pinpoint new aero parts on the cars, discussing what technology was in play or their purpose. And that became *Development Corner*, named simply because it was about developments and was presented in a quiet and dimly lit corner of the Barcelona media centre.

Finally, let me explain about the cheese. I have a fondness (fondue-ness?) for using cheese to illustrate complicated technical developments within F1, because it just works. This began in Bahrain in 2020, when the FIA removed a triangular section of the back of all the cars' floors to try and cut some of the downforce. It didn't seem like a big change, but it was in a crucial area of the car, and when I looked at the bit they cut out I thought it resembled a triangular slab of Swiss cheese. So on the way to the track I bought some Gruyère, used a ruler and knife to cut it to the exact dimensions of the bodywork in question, and took it into the paddock to hold up to camera and explain the rule change to the viewers.

A little later I turned up to an interview with Sebastian Vettel, still with this triangle of cheese in my carrier bag. We were familiar enough that I knew he'd say, 'What's in the bag?,' which was a perfect opportunity for me to ask if he wanted some of my cheese. Taking this evidence of my eccentricity in his stride, he added, 'What are you going to do with that?' I said, 'Well, I can't do anything. I can't eat it, it's been sweating in this plastic bag in the Bahrain sun all day.' The cheese had gone all wobbly, so Seb had a go at me for being wasteful!

When we got to the ground-effect era in 2022, in order to explain the downforce-generating venturi channels underneath the cars I used a block of Cheddar, and carved out channels in the correct dimensions. On another occasion I thought the sidepod scoops on the Ferrari looked like the semi-circular wax casing of a Mini Babybel. So I put that to Carlos Sainz, who looked a little confused when I held up an example that I'd bought in a Barcelona supermarket – although after a demonstration, he had to agree.

I try not to keep too strictly to the format of *Ted's Notebook*, especially after 14 years. I like to shake it up now and then, and every year we do something a little bit different. In 2025 we're doing a *Learn with Lewis* section (when I remember to do it), where we acquire a new Italian phrase every week. He's learning Italian at Ferrari, so we're going to learn it with him. I'm looking forward to finding out how to say things like, 'Can we change the rear anti roll bar?'

Twenty-four years of what I'd like to hope are insightful contributions to practice, qualifying and race coverage, breaking stories and painstakingly crafting features, and yet I'm best known for a rambling, one-person, one-shot, live monologue – a programme I didn't think of, didn't name, reading off a notebook I didn't even pay for. However, if it is informing and entertaining the audience, I'm happy. Even if they are only watching to see Pete and me stack it on an escalator.

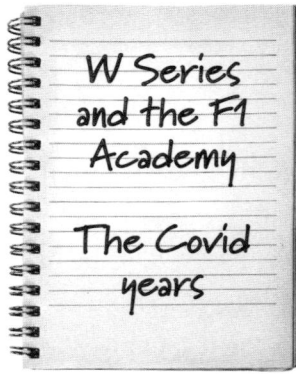

W Series
and the F1
Academy

The Covid
years

Chapter 19

Surviving, Driving

———

The Nag's Head, Covent Garden. A traditional English pub. Not your usual venue for an F1 sponsor day, but then this wasn't your usual sponsor. Organized by experienced motorsport PR Lizzie Brooks, it was hosted on a Monday in early March 2020 by a company called ROKiT, which has commercial interests in mobile phones, motorbike cleaning products, tequila and beer – hence the pub.

ROKiT chairman Jonathan Kendrick was there to talk about the company's upcoming sponsorship of the Williams F1 team, the Venturi Formula E team and the all-female W Series, which was just beginning its second season. 'JK' as he's known is a larger-than-life character. However, Lizzie had spent two years working for Eddie Jordan, so knew how to handle a character. Team principal Claire Williams was there with her former driver Felipe Massa (then at Venturi Formula E). Also present was Jamie Chadwick, Williams's development driver and inaugural champion of the W Series (the world's first all-female racing series) and the CEO of that championship, Catherine Bond Muir.

Mini hamburgers were served at the back of the room, and we were given a particularly tasty German beer called ABK, one of the ROKiT brands, from a town southwest of Munich that I'd never heard of. I've since struggled to find ABK, which is a shame as I was quite taken with it, although only one of the three organizations present that day retained ROKiT's sponsorship, which indicated that the company may have had bigger issues on their hands than beer distribution.

I had been involved with the W Series since its debut in 2019, having found myself temporarily looking for work. Two of the questions I'm asked most, and that I used to ask myself, are, 'Why haven't there been more female F1 drivers?' and 'Will there be one in the next decade?' There is objectively no reason why there couldn't be a successful female F1 driver. Motor racing is one of the very few sports in the world in which women and men compete on equal terms. On average women are smaller and lighter than men (always a benefit in a racing car where mass slows you down), studies have shown that they process information faster and sports such as ultra running have proved that women can outstrip their male counterparts when it comes to endurance. Why then can I count on one hand the number of women who have driven in Formula 1? What has been holding them back? W Series was incredibly important, because while it might not have had all the answers, it did remove three of the barriers that were preventing female drivers from getting into single-seater motorsport.

The first was money. Based on the existing status-quo within the sport, most sponsors didn't think female drivers were likely to make it into F1, so a frequent hurdle was trying to raise a racing budget. W Series turned that on its head. The way the series was funded meant that all a driver had to pay for was their travel to their nearest major airport – W Series took care of the rest. On top of that, the drivers were actually paid, unheard of in the junior formulae. There would be prize money for everyone, and the championship winner received $500,000 with which to potentially fund their next drive.

Second was getting practice time in the cars. Any testing outside of a race weekend costs money, and these were not drivers with Formula 1 teams backing them up. The only way any driver improves is by practising, and W Series gave its drivers plenty of testing so they could sharpen their skills and compete at the higher levels. Thirdly, by being open, visible and publicizing the fact there was a series just for female drivers, W Series inspired a whole new generation of girls to see motor racing as something they could do, something that was open for them to try.

W Series's first season was held on the German Touring Car Championship support bill (anyone who has attended a major race will be familiar with the support races that keep the crowd entertained until the main event), and I saw first-hand how girls who came with little enthusiasm to Hockenheim or the Norisring with their families were suddenly entranced and inspired by these incredible female racers.

Surprisingly, there was some criticism of the series from a few female racing drivers including IndyCar drivers Danica Patrick and Pippa Mann, who felt W Series made it harder for female drivers to compete equally with men if they were segregated into a separate championship. But I would argue that there were so few female drivers coming through, and the barriers were so clear, that it seemed obvious some kind of positive action was needed to create new opportunities and broaden accessibility.

It was rare to find a startup that had such a strong group of people behind it. Founder Catherine Bond Muir had a decade's experience in intellectual property law and corporate finance. David Coulthard was on board, as was his friend Sean Wadsworth, chairman and cornerstone investor. Also involved were my old friend Dave Ryan from McLaren and Matt Bishop, former McLaren communications director, who had joined in a senior comms role because he believed in the project and liked working with good people. Lee McKenzie presented the broadcast coverage, I was the roving reporter, and Claire Cottingham was joined by Coulthard in the commentary box.

They came from a wide range of backgrounds with differing levels of experience, but the drivers were the stars. Alice Powell had been the first woman to win the UK Formula Renault championship and the first to score points in the GP3 series, but then she ran out of money. It was a similar story for Sarah Moore – she had won the Ginetta Junior Championship in 2009 (the same series in which five years later Lando Norris would finish third) but couldn't find the funding needed to move up the ladder. Other drivers like Naomi Schiff and Jessica Hawkins had more experience in GT cars but hadn't until this point been given a chance to drive many single-seaters. What W Series gave them and around 40 others who raced in it over three seasons was the opportunity to compete, and to showcase their skill and dedication without having to worry about where the next sponsor was coming from.

Finding that sponsorship money was for W Series to worry about. The organization attracted some great partners over the years, including Puma, Heineken and my pub hosts ROKiT, but rising costs and a year lost to the pandemic meant that by the end of its third season in 2022 W Series badly needed to find a new injection of money. By now it had joined the F1 travelling circus as a support racing series. While that was fantastic for prestige, publicity and the patronage of a few F1 teams, having to ship the cars all over the world was costing them millions.

At the start of 2023 Catherine Bond Muir's team had a new investor lined up, but at the last minute, unexpectedly, they withdrew, and she had to take the gut-wrenching decision to lay off staff and take the company into administration. There had been rumours of an opportunity around 18 months earlier for Sean Wadsworth and his fellow investors to sell the company to Liberty Media, owners of F1, but a deal never materialized. It wouldn't have taken much to keep the series alive – an F1 team's marketing budget would have done it – but time ran out and Bond Muir was forced to admit defeat.

What W Series did achieve was to prove beyond doubt that the concept worked, but that maybe a different financial model was required. Liberty Media certainly thought so, and asked Susie Wolff, the last woman to drive a Grand Prix car in a race weekend, to head up a new all-female championship by the name of F1 Academy. The drivers would have to pay €150,000 per season to compete, but that would be matched by Liberty. The existing Formula 2 teams would run the Formula 4-spec cars and the F1 teams got involved by supporting a driver each. Lia Block, for example, was selected to the Williams Driver Academy, so she races in Williams colours. But this system doesn't cover all the drivers, some of whom race instead in consumer brand cars representing the series sponsors, for example Aurelia Nobels, who is supported by Puma. Other sponsors such as American Express, Charlotte Tilbury and Tag Heuer keep the money coming in.

Having been involved with getting W Series established on TV, I'm proud that its legacy lives on and we show every F1 Academy race live on Sky Sports F1. Wolff and her team are doing an amazing job in continuing to break down barriers and it's exciting to see some seriously talented drivers coming through. Abbi Pulling dominated the 2024 F1 Academy season and now races Formula 3 level cars in the British GB3 championship. Whether it's her or someone yet to come, it wouldn't surprise me to find a female driver racing full-time in F1 in the next decade.

On that day in 2020 at The Nag's Head, however, all Catherine Bond Muir and Jamie Chadwick were hoping was that their second season would get going without a hitch. Lizzie Brooks called the room to order; I popped my plate on the side and sat down to hear what everyone had to say. Jonathan Kendrick talked about his pleasure in sponsoring Williams after working for Sir Frank Williams as a tyre man in the 1980s. There were a couple of questions from the floor about the upcoming F1 season and to Felipe Massa on how he was enjoying his move to Formula E. Then *The Sun*'s F1 correspondent Ben Hunt piped up with a question for Claire Williams. 'There's a lot of talk about this virus that's coming out of China, and that's in Italy now. How concerned are you that it could affect sponsorships and sporting events with a lot of people gathering together, like at the Australian GP?'

'What a question,' I thought to myself. 'Some virus is hardly going to stop F1 from going ahead, is it? He must have a story he's trying to pad out with quotes.' Williams gave a straightforward answer about how everyone was keeping an eye on developments and trusted the FIA and F1 to make decisions about the calendar.

The press meeting wrapped up, I said my thanks and farewells, and mentioned to Jamie and Catherine how much I was looking forward to getting going with W Series for our second season. As I walked out of The Nag's Head into the spring sunshine I stepped over a floor mosaic by the entrance that read 'Established 1827.' 'This pub has been here for nearly 200 years,' I thought to myself. 'It's survived two world wars, a couple of global recessions, and whatever Covent Garden throws at it every Saturday night. A virus isn't going to hurt it.'

A few days later I flew to Australia via Singapore. I was particularly excited because I had booked a Star Alliance 'round the world' ticket with my return on United Airlines, flying across the Pacific from Melbourne to San Francisco. I was hoping to have seen Hawaii from the air before connecting on to a San Francisco to London flight to complete my circumnavigation of the globe. It was a longer route, but it appealed to my sense of adventure, as much as sitting in comfort on an aircraft can be adventurous. On landing, we'd seen the latest news about the virus – it already had a name, 'Covid-19', but on the Wednesday before the race, it had a classification: Pandemic.

Wandering around Melbourne, you wouldn't have known anything was different. Then that evening, news emerged that some members of the McLaren and Haas teams had displayed 'flu-like' symptoms on arrival in Australia, and had been quarantined in their hotel rooms and tested. With their results pending, the F1 paddock looked different when we arrived on Thursday for media day. Guenther Steiner was the first interview of the day, and he sat in a corner of the team's paddock area with a table in front of him. Journalists placed their recorders on the table, behind which was a one-metre exclusion zone enforced by a retractable queue barrier. Behind that sat all the journalists, together. It was very odd. It seemed that the teams were afraid that the journalists would pass on the virus to the team members, when in fact, given the first reports of cases had been on the teams' side, the concern should have been the other way around.

The interviews continued throughout the day. At one point, I measured my arm to see if it was a metre long, so as to ascertain if I could hold a microphone close enough to any driver we were being socially distanced from. Everyone I spoke to had been pretty non-committal about whether they thought the race would be affected by the pandemic. Most wanted to fall back on the comfortable ground of lap times, setups and the FIA's recent response to questions surrounding Ferrari's engine.

Then came a bombshell. In the FIA press conference Lewis Hamilton, F1's biggest worldwide personality, had the courage to say what many were starting to think. 'I am really very, very surprised that we're here. I think it's great that we have races but for me it's shocking that we're all sitting in this room.' Hamilton added that he'd seen what the rest of us had in Australia, life going on as usual, and was concerned about how many people were gathered together in one place for the Grand Prix.

In an instant, the assembled journalists knew they had a massive story on their hands. A well-worded follow-up asked Hamilton, 'Why are we still here?', to which he replied, 'Cash is king. Honestly, I don't know. I can't really add much more to it.

I don't feel like I should shy away from my opinion. The fact is we are here, and I just urge everyone to be as careful as you can be.' I'm not sure if Lewis meant the phrase 'cash is king' to be a throwaway line, but it made headlines around the world, hinting that as long as money was being generated by F1 for the teams and the circuits, it didn't matter about people's health. It was controversial but would be proved accurate in that when the money did stop flowing during the subsequent lockdowns, several F1 teams almost folded because of the lack of cashflow.

While all this was happening in the media centre, the team members' test results came in. The Haas mechanics only had common colds, but McLaren's test returned positive. F1 hadn't yet established procedures to follow in the event of a positive test, so when it did happen, McLaren suddenly had to isolate employees in a bid to prevent further transmission. As more and more people were identified as close contacts and were sent back to their hotel rooms, it quickly became clear to McLaren management that they didn't have enough people to crew their racing team. On Thursday evening the team announced that it was withdrawing from the Australian GP weekend. It was a big shock, and even people who thought they knew what was going on inside the paddock had no clue, such was the pace at which decisions were being made.

That evening a crisis meeting was held. The word 'stakeholders' is often overused in a mist of management speak, but it seemed particularly apt for this assembly of people at Melbourne's Crown Casino Hotel. In addition to the F1 team principals, the top people from Formula One Management, the FIA and the AGPC, the Australian Grand Prix Corporation, were joined on conference calls by the local police, Victoria state health officials, lawyers and local government officials. Time was not on their side. Track inspections started at 7.40am and the gates to Albert Park were due to open at 8am. It was a tense meeting, and the financial stakes were high given how much money had already been spent and contractually promised. The only teams who committed to continuing with the Grand Prix and racing with or without McLaren were Red Bull, sister team Alpha Tauri and Racing Point.

Mercedes was diametrically opposed. On Thursday evening Mercedes wrote to F1 and the FIA requesting the cancellation of the race, reasoning that the team could not guarantee the safety of its employees if the event continued, it wasn't right to race if McLaren could not, while empathizing with the worsening situation in Europe, especially in Italy. Mercedes concluded by saying that it would pack up its equipment in the morning.

Clearly the situation was now turning legal. Mercedes couldn't say outright that it was withdrawing from the event, as it had legal obligations to continue, but that's effectively what the team was doing. Ferrari was equally very concerned about the state of affairs back home and was as keen as Mercedes to fly back before Italy locked down.

The team bosses argued about the legal and sporting aspects of the situation into the evening. Managing director of motorsport for F1, Ross Brawn, was in and out of the meeting as decisions swayed from carrying on to cancellation and back again. The organizers even considered continuing with local support events only, but abandoned that plan when it was pointed out that fans would not be too happy watching a touring car race while the F1 teams packed up their garages.

Finally, at around 1am and following a day of further meetings in Europe, a decision was made. The race would be cancelled. Matters were left in the hands of the local authorities and lawyers to argue over liabilities. Ferrari's team principal Mattia Binotto could see that the race wouldn't take place, so called his drivers Sebastian Vettel and Kimi Räikkönen to tell them to get on the first flight out of Melbourne. They were only too happy to get back to their families, so packed their bags and arrived at Melbourne Airport at 4.45am for the 6.15am Emirates flight 409 to Dubai. Among the chaos and confusion, this was the clearest sign that the race was off. A 'whistle blower' posted a photo on social media of the flight manifest, printed, as they all still are, on a dot-matrix printer at the airport gate. The seat plan for flight EK409, on Friday 13 March 2020, was as follows: In 1A (with an Emirates gold card), Mr Sebastian Vettel, and in seat 3A (also with a gold card and a gluten-free meal request), Mr Kimi-Matias Räikkönen. There was the usual doubt when dealing with anything on the internet as to whether this was a fake, but as soon as I saw the dot-matrix detail and the use of Kimi's passport name of Kimi-Matias, I knew it was the real thing.

While we were sure that Seb and Kimi wouldn't be driving in the Australian GP, joining their McLaren and Mercedes colleagues Lando Norris, Carlos Sainz, Valtteri Bottas and Lewis Hamilton, the remaining 14 drivers woke up on Friday morning not knowing whether to head for the circuit or the airport. They were in good company. Tens of thousands of people turned up but weren't allowed into Albert Park. I had seen the Vettel/Räikkönen flight manifest picture over breakfast, which told me all I needed to know about the fate of the Grand Prix, but we were still at work, and although the gates were closed to the fans, the teams, media and officials were being allowed in. I'll never forget how bad I felt walking past a huge queue of ticket holders as we went through the gates. Some were asking us if we knew what was going on, and I answered as honestly as I could that it didn't look good, but there was still no official confirmation either way. The picture had been confused still further by various health and governmental agencies in the state of Victoria not wanting to be the ones who shut down the Grand Prix and offering last-minute solutions that were unlikely to work.

Mercifully the fans didn't have to wait outside too long. Around three hours after Seb and Kimi had taken off for Dubai, F1, the FIA and the AGPC announced the cancellation of the event. There was talk of a postponement, but this seemed unlikely when fans were offered a refund on their tickets. We were in our morning meeting in the TV compound when the news officially dropped. It was a real 'wow' moment. It wasn't what we came to Australia for, but we had a programme to make. For reasons I didn't understand, rather than stay at the track where the story was, our team was sent off-site to present a special programme from a café outside our hotel. As everyone else schlepped cameras, lights, tripods and sound equipment back across Albert Park, I put my radio mic kit on and headed for the paddock with Lee the marine.

Teams were already halfway through their pack-up procedures, unused fuel was being returned, tyres were being wheeled back to Pirelli. Equipment was being packed away with a bit more care than usual given the mechanics weren't sure when they were going to see it again. We filmed some of this activity for posterity before relocating to just outside the paddock entrance for probably the strangest press conference I'd ever attended. Then F1 CEO Chase Carey was joined by the AGPC chair Paul Little, its CEO Andrew Westacott, and FIA race director Michael Masi. They explained the situation and why they were left with no option but to cancel the event, stressing how sorry they were for the fans. Carey was asked for his response to Hamilton's suggestion that F1 had only travelled to Australia in a pandemic because 'cash is king'. 'I guess if cash was king, we wouldn't have made the decision we did today,' he replied. A fair response, but in reality, there was no alternative. Borders were being closed all over Europe, and it was only a matter of time before Australia did the same.

I broadcast a couple of reports from outside the paddock, one about what was going on with the teams and drivers and another reflecting on the news lines from the press conference. Simon Lazenby wrapped up our programme from the other side of the park, said goodbye to our viewers, and we went back to the TV compound to pack up our stuff. It was the strangest feeling knowing that the season had not even started but we were saying, 'Goodbye' and 'See you when all this blows over.' Then it was back to the hotel and try to figure out how to get home. The next day, while Helen Cox, Nick Warren and their team at Travel Places worked feverishly trying to find ways of getting all the UK-based teams and media home a few days early, the crew and I went for a very pleasant Saturday lunch on St Kilda Beach. There was a moment later that afternoon when we were in the restaurant's bar and the local news was on the TV. Suddenly it cut to one of my reports from the previous day about what the F1 teams would do next, and I got a small cheer from the locals who were wondering why the guy on the screen was now standing there in front of them with a glass of the Yarra Valley's finest in his hand.

The last we'd see of each other was Sunday morning. My plan to fly around the world hit the buffers as the USA was starting to close its borders and there wasn't an assurance that when I got to San Francisco I'd be able to fly on to London. Pleasant as California is, I didn't want to be stranded there away from my family for the next three months, so flew back on Qatar Airways via Doha (indeed Qatar Airways never stopped flying during the Covid pandemic, providing an important lifeline to many). Two days after we got back to London, the UK closed its borders, told its population to stay at home, and that was it. I should have listened to Ben Hunt at The Nag's Head.

From breadmaking to online workouts, everyone went a bit stir-crazy during lockdown, but as a media company, we needed to figure out a way to meet the increased demand for our content. After a few weeks we pivoted to making programmes remotely and attempted to keep our audience informed and entertained by creating shows from our homes. We did 'watch-alongs' of classic races featuring the drivers who took part in them, a series called 'At home with . . .', where each of us on the Sky F1 team introduced three of our favourite features that had aired over the years, and told some stories about how they came to be, what happened during filming and so on. We filled out the schedule with extended highlights of previous races.

Things were still happening in F1. Sebastian Vettel split from Ferrari and was later confirmed at Aston Martin to replace Sergio Pérez, and there was worrying news from teams who were struggling to find ways of paying their staff, given that there was no appearance money coming in from the F1 circus. As soon as we were allowed out of our homes for a daily period of exercise, I offered my producers a weekly edition of *Ted's Notebook*, to round up all the bits and pieces of news I'd heard about. Given social distancing, I had to be my own cameraman, so I found a way to clamp my little pocket motion-stabilizing camera on to the end of a selfie stick, which seemed to work reasonably well as a technical solution to the problem of how to film myself while also walking round my local park. A couple of laps were enough to fill viewers in on the latest in F1 and what plans were afoot for the return to racing. I would never have imagined presenting *Ted's Lockdown Notebook*, and I sincerely hope I never do again, but in the absence of anything else to watch the shows were fairly popular and I still get people coming up to me from time to time saying how much they enjoyed them.

I'd like to think it was my lockdown rambles, but of course the thing that really skyrocketed F1's popularity during Covid was a new documentary series from Netflix called *Drive to Survive*. Commissioned by a television man, former ESPN executive Sean Bratches who moved to F1, the first season had aired with reasonable success but hadn't set the world alight. A second season had been commissioned, though, and the documentary makers had been out at the races filming all year. It was released worldwide on Netflix at the beginning of March 2020 and became a must-watch for people stuck at home, huge numbers of whom had never heard of or watched F1 before. They grew to love the drivers and team bosses through the stories showcased in such a vivid and compelling way. Many of these new viewers were female and from younger age groups, two demographics that F1 had previously struggled to attract.

Much of this new audience was also in the USA and it's no exaggeration to say that, after decades trying to break the American market, F1 had *Drive to Survive* to thank for finally achieving it. It seems incredible now, but in the first season Mercedes and Ferrari had been reluctant to let the cameras go behind the scenes, and thus didn't allow the Netflix crews to fully embed themselves for a weekend. When they saw the final result and it dawned on them that it would be of benefit to their sponsors to be seen by such huge numbers of viewers, especially in the US, they opened their doors.

One of the early stars who came to worldwide fame thanks to *Drive to Survive* was the entertainingly sweary Guenther Steiner, at the time the team principal of the American-owned Haas F1 team. A favourite episode showed Guenther in his natural habitat, perched on his pit-wall command post. 'F**k!' Steiner barks at his monitors on the pit wall, watching his driver Romain Grosjean make a mistake in qualifying. 'I'm f**king speechless, I can't f**king believe it.' Grosjean is getting it in the neck and Steiner admits to the then chief engineer (and now team boss) Ayao Komatsu that his patience is wearing thin. 'Ach, f**k me. I can't keep finding excuses for him.' After hearing a string of Grosjean's own excuses for another race out of the points, Guenther responds, 'Just focus on driving, not f**king whinge.' Another episode shows Steiner hosting a team dinner at the French GP – Grosjean was absent. 'Maybe I didn't invite him because he doesn't deserve any food.'

It was deliciously entertaining stuff. And refreshing to see so much passion from an F1 team boss. Impossible to know if Steiner forgot he was being followed around by a camera and a microphone recording his every word, or just didn't care? The Netflix series made him famous, and even foretold his own departure from F1 when he explained that everyone is accountable and if the team doesn't perform, the buck stops with him. Steiner was relieved of his Haas duties on the eve of the 2024 season.

Like everyone, I love watching the episodes, but it's interesting to consider how they are put together, without many of the constraints in covering a race weekend that we are bound by at Sky. Firstly, the *DTS* cameras are everywhere and they film absolutely everything. There's a good reason for that – when the season starts the producers don't know what the stories are going to be. Midway through the season they're in a position to see what the emerging lead stories are, and what might make for a good episode, but until that point, you have to have the shots in the can. Secondly, Netflix are allowed in everywhere because the teams know that the footage won't be seen until the beginning of the following season, by which time it will be old news and not as sensitive. Final approval of *Drive to Survive* rests with Netflix and F1, but the teams and drivers are shown their sections ahead of broadcast and can make representations to the producers.

Thirdly, and media students will be writing theses on this well into the future, consider how the presence of the *Drive to Survive* cameras is affecting actual decisions being made in the F1 teams. Are events being created or staged when the cameras are rolling because the teams know it will guarantee them making it into an episode, giving their sponsors millions of eyeballs on this worldwide entertainment platform? What if there's something juicy about to happen but the cameras are somewhere else, or not present at all? After all, the *Drive to Survive* camera and sound crews don't live with the teams and drivers 24 hours a day – they have to be booked for a day's filming like any other service. Think about that process – would it start with a call from the team's press office to the *DTS* producers offering them a certain filming opportunity on the promise that something important is going to happen? Would that have been happening anyway or is it only happening in that way for the benefit of the cameras? And the team bosses? How have fame and recognition affected them and their decision-making process?

Finally, consider how much footage and how many behind-the-scenes stories didn't make the final edit, and what might happen to that material in the future. It's a fascinating thought that the cameras might have caught something that was too spicy or sensitive even for *Drive to Survive*, but which might be declassified and broadcast in 10 years when it's safe to do so. I'd watch that!

That race

Last lap drama

Chapter 20
The 2021 Abu Dhabi GP

If you've read up to this point and been wondering when I was going to get to perhaps the most significant and important moment of my time in F1, and certainly the biggest story I've covered, well, here it is – the Abu Dhabi Grand Prix of 2021. It has become known as one of the most controversial moments in the sport's history, but since the facts of what happened that evening are known and generally accepted, why the controversy? No one disputes the course of events of this championship-deciding race, most of it was broadcast live and the players have all since had their say.

What is disputed is whether the result was fair – impartial and just, without favouritism or discrimination – and everyone is entitled to their opinion based on the evidence and facts. Of course, there is a feeling of injustice on one side and a defensive position on the other, all perfectly natural and understandable emotions. However, there's something about this race that makes people unwilling to even talk about it. Indeed, it would probably make my life easier not to write this chapter at all, for fear that the mere re-telling upsets one side or the other. However, if the study of history is to better understand, empathize and learn lessons for the future, and if you're interested in hearing the facts of what happened from someone who was there in the pits that evening and witnessed events unfold first-hand, here we go.

The story starts a couple of years earlier with the tragic death of veteran FIA race director Charlie Whiting. It was on the Thursday morning of the 2019 Australian Grand Prix that we learned the news that Whiting had died in his hotel room overnight at the age of just 66. It was a shocking occurrence, one of those events that sends the paddock into a very subdued place. Charlie's passing was the only thing that anybody was talking about on that day.

He had been a part of the F1 scene for over four decades, initially as a mechanic for Hesketh, and then for Bernie Ecclestone's Brabham team. He moved to the FIA in 1988, and over the years his role expanded to encompass a range of job titles alongside that of race director, including official starter, safety delegate and head of the technical department. A workaholic and perfectionist, he commanded respect across the paddock. Drivers and teams didn't always agree with his decisions, but they accepted them. Whoever you spoke to that day in the Albert Park paddock, be it mechanics, drivers, team bosses or journalists, everyone was shocked, especially as he'd been busy at the track with his usual pre-weekend duties just the previous day.

At the same time as dealing with the shock and making appropriate arrangements for Charlie's family, the FIA had to address how the race meeting could progress, given that he held such a crucial position. Few knew about it, but Whiting had been training up deputies in the race director's role. Australian Michael

Masi had years of experience gained in the hectic world of the V8 touring car series in his home country and had been appointed as a deputy race director by the FIA in 2018.

The FIA issued a document naming Masi race director for the Australian GP. Initially engaged on a race-by-race basis, he was eventually awarded the job full-time. Replacing Whiting was an incredibly difficult task, and considering the scale of the challenge, Masi started relatively well. He played himself in, got to know the drivers and the teams, and worked to get F1 back on track during the Covid-19 pandemic in 2020.

During his tenure as race director, Whiting had agreed to requests from Formula 1 to grant their cameras more access to the rule-making and rule-enforcing side of the sport. F1 TV cameras were subsequently allowed into selected Friday evening drivers' briefings, which produced fascinating insights. Charlie had also agreed to hold post-race media briefings to discuss decisions he'd made during the weekend. After a period getting used to his new job, FIA officials began to make Masi available to the media, too. This both raised his profile and afforded much more scrutiny of his actions.

There was also a desire to make the mid-race decision-making process more transparent, so F1 asked for intercom messages between Masi in race control and the teams on the pit wall to be made available for broadcast – something I can't imagine Charlie agreeing to. This was new for F1, but a high degree of transparency was already common in sports such as rugby, cricket and tennis, where the referees or umpires routinely wear microphones so their decisions can be heard, and it was broadly accepted that it was good for F1 to follow suit.

There had long been an intercom channel between the teams and race control. Often the chat was about routine or mundane matters, such as why a session had been delayed or whether an extra formation lap would be needed. At other times the teams would report their rivals to the race director if they felt a rule had been broken. Whiting always took these appeals with a pinch of salt – it was part of the game, and he had a measured approach when it came to deciding which messages to act on. Charlie would hear mainly from team managers over the intercom, although team principals could ask questions and indeed did start doing so more often when, under Masi's tenure, the messages began being broadcast.

With all this change afoot, Michael Masi started to learn who he could lean on for support and advice. In Red Bull Racing's sporting director Jonathan Wheatley, he found a valued colleague and the two forged a professional relationship. Wheatley was one of the pit lane's leading experts on the rules, given that he'd been responsible for them at Red Bull since 2006. It was not only Wheatley – other team sporting directors such as Ron Meadows of Mercedes, Alan Permane of Renault/Alpine and Andy Stevenson of Racing Point/Aston Martin were also important sounding boards for Masi, as all had worked for many years with Whiting and knew the rule book by heart. However, I often saw Wheatley taking time to discuss matters with Masi, and he could be heard on the broadcasts taking a more collaborative, proactive tone with Masi than some of his fellow sporting directors.

One example of Masi showing that he was happy to accept Wheatley's advice occurred at the 2021 Azerbaijan GP when Max Verstappen's Red Bull crashed on the pit straight. The immediate cause appeared to be a rear-tyre failure. During the ensuing safety-car period Wheatley suggested that a red flag might be the best solution to allow everybody to stop racing and inspect or change their tyres, because Verstappen's crash had been so unexpected and had obvious safety implications. Lance Stroll had also suffered a suspected tyre failure earlier in the race. Wheatley's suggestion was timely, as Masi later said that he had already been thinking of stopping the race and allowing everyone to fit new tyres. The race restarted, and Red Bull went on to win with Sergio Pérez.

That 2021 season was Masi's third in charge, and it developed into an intense battle between Verstappen and Hamilton. There were plenty of dramatic incidents along the way, clashes between the two such as at Silverstone, when they battled for position into Copse Corner, which led to Verstappen having a huge crash and needing a trip to the hospital. Hamilton was given a 10-second penalty for causing the accident, which he overcame, going on to win the race. Mercedes's victory celebrations that day went down very badly with Red Bull and the Verstappen family, who felt them insensitive and disrespectful. Later in the season at Monza, Max and Lewis went into Turn 1 side by side. Neither would give ground and their wheels interlocked, flipping the Red Bull up to land on top of Hamilton's Mercedes. Verstappen was given a three-place grid penalty for the following race after the stewards judged he was to blame for causing the collision.

As the end of the season approached, Hamilton scored consecutive wins in Brazil, Qatar and Saudi Arabia, thanks to some late-season improvements that made his Mercedes significantly quicker than the Red Bull, but tensions ramped up between teams even further. In Brazil, as Hamilton was trying to overtake on the outside of Turn 4, Verstappen widened his line and eased the Mercedes off track. Not only was this permitted by the driving guidelines at the time, but Michael Masi had also committed to a general philosophy of 'let them race' whereby he preferred to allow the drivers to race each other hard, rather than referring every strongly defended move to the stewards.

Laudable though this aim was, it resulted in chaos, particularly in Jeddah a week before Abu Dhabi, when Verstappen was penalized twice, once for causing a collision with Hamilton and then for having left the track and gained an advantage. Hamilton declared Verstappen's racing to be 'over the limit' that evening while Max complained that 'F1 is more about penalties than racing'.

Hamilton's win in Saudi Arabia meant we headed into the final weekend in Abu Dhabi with the two title contenders on 369.5 points apiece. However, Verstappen had scored nine wins to Hamilton's eight, so if neither man scored in Abu Dhabi, the Dutchman would win the title.

Throughout practice the Mercedes was clearly quicker, and it was only a brilliant lap from Verstappen in the final part of qualifying that clinched him pole position. But on recent form everybody expected that Hamilton would have the faster car in the race and would be able to somehow get to the front and finish ahead. I remember sizing up the contenders in my *Notebook* on Saturday. I went up to random people in the paddock, and just asked, 'Hamilton or Verstappen?', and they gave their reasons as to which driver they thought was going to win the world championship. Certainly pace-wise it looked like Mercedes had the edge, but none of us could have predicted what a memorable race it would turn out to be.

The drama began on the first lap. Hamilton got the better start and led into Turn 1, but Verstappen got a run on him down to the Turn 7 chicane. The Red Bull went for a pass down the inside and, like Brazil, walked the Mercedes to the outside kerb. Rather than concede the corner, Lewis drove across the run-off and rejoined the track ahead of Max. In the context of what happened in Jeddah, I expected Hamilton to voluntarily give the position back, looking to avoid a potential time penalty. As Verstappen was voicing his objection on the radio, Wheatley was already pleading his case to Masi, who referred the incident to the stewards, but they concluded that because Hamilton had been given no room and had reinstated the original gap to Verstappen, no further investigation was necessary. There was a radio message where Max was told that there would be no action, and he said, 'No surprise there.' In his eyes, he had been punished the previous week in Saudi for the same thing Hamilton had gotten away with now.

With better tyre usage and quicker pace Hamilton pulled away, but in the pit-stop phase, fell victim to a pre-planned Red Bull trap. When Hamilton rejoined following his pit stop on lap 14, Verstappen's teammate Sergio Pérez had been instructed to hold the Mercedes up for as long as he could. He managed to do so for nearly a lap before Lewis could get by, which, crucially, cost Hamilton nearly eight seconds. Max later called Sergio 'a legend' as he thanked him for his help. The time loss allowed Verstappen to close in on Hamilton. It wasn't the end of the world for Mercedes, they were still ahead and still quicker, but it did mean that in the event of a safety car, Lewis did now not have the handy 15-second time gap that would allow him to make a pit stop and rejoin ahead of Max. As long as there wasn't a late-race safety car, he'd be fine.

On lap 52, with just six laps left to run, Lewis was over 11 seconds clear. An eighth world championship was in his sights.

Further back in the race, Haas's Mick Schumacher was having a spirited fight with Williams's Nicholas Latifi for 15th place. Schumacher ran Latifi wide at Turn 9, which got the Williams's tyres dirty as he left the track. With poor grip, Latifi lost control of his car and crashed into the barrier on the exit of Turn 14, the left-hander after the circuit passes under the glitzy Yas Hotel. There's no run-off at that point, and with the damaged car stranded on the track, Masi was forced to neutralize the race by deploying the safety car, allowing marshals to clear up the heavily damaged Williams. The worst possible scenario for Lewis Hamilton and Mercedes had just become a reality.

It's a quirk of late-race safety-car periods that they often disadvantage the driver in the lead, because the second-placed man can react to what the leader does and do the opposite. Mercedes could see that the Red Bull mechanics were out in the pit lane with fresh new tyres for Verstappen, but that didn't mean Max was definitely going to pit. Had Mercedes brought Hamilton in, Red Bull would (probably) have told Max to stay out, which would have given him the lead of the race. What neither team could know was how long it would take to clear the Latifi crash, and whether the race would end before all the things that the rules said had to happen could happen before racing resumed. If the race finished under the safety car and Mercedes had handed Max the lead (and thus the world championship) by pitting Hamilton, the team would have looked pretty silly.

The Mercedes strategists led by James Vowles were under pressure and running out of time. Thanks to Sergio Pérez, they did not have the luxury of pitting and staying ahead of Verstappen. It looked like quite a big crash, so might take a while to clear up. Six laps left. They made the call – stay out. As the second car, inevitably Red Bull did the opposite with Verstappen, and brought him in for new soft tyres. It was the easiest decision in the world for Red Bull's strategists Will Courtenay and Hannah Schmitz. Verstappen was so far ahead of the third-placed Ferrari of Carlos Sainz that he didn't even lose his second position – it was a free stop.

When Verstappen came out of the pits there were five lapped cars between him and Hamilton. They had stayed out, knowing that the way the safety-car rules were usually enforced would allow them to unlap themselves, leading to a better result at the end. This is a crucial point. Under a safety car Masi had the option not to allow lapped cars to overtake the leaders and unlap themselves. It was at his discretion – he could decide that everyone should hold station if he felt the unlapping procedure would use up too much time. Masi had done this before, at the final safety-car period of the 2019 Brazilian GP. Recovering Latifi's car took longer than was perhaps initially expected, with a brake fire occupying the marshals for a few crucial minutes. Getting the fire extinguished and the car removed was Masi's priority. Sorting out how to finish the race would come later.

It's worth noting that Masi did have the option of stopping the race by showing the red flag, which would have allowed the marshals to take their time. Under a red flag everybody can change their tyres, after which there would be a restart for the remaining laps with everyone on new tyres and on equal terms. This would have been the fairest way to decide the race. That's what happened in Azerbaijan earlier that year, when we had a late stoppage and what was in effect a one-lap race after the restart. However, there was a consensus among team bosses in the F1 Commission (the body that, among other things, debates the rules) that there had been chaos at the restart on that day in Baku, so they would rather continue with the safety car if possible. Knowing this, Masi opted to continue under the safety car in the hope that the race would be able to resume.

Once he'd made that decision, the longer the marshals took to shift Latifi's wrecked car and sweep up, the safer the Mercedes call not to have pitted Hamilton looked. The team was banking on the remaining laps running out, allowing Lewis to finish the race in front. Red Bull, meanwhile, needed a restart and at least one racing lap to give Verstappen a chance to make use of his new tyres and get ahead of Hamilton. On the Red Bull pit wall, Wheatley and his boss Christian Horner were running out of time.

Masi's initial decision, which came up on the pit wall timing screens, was not to allow lapped cars to overtake, as he had done in Brazil two years previously. The intention behind this call was to avoid the time-consuming unlapping procedure which would help squeeze in one or two racing laps. This was Masi's preferred conclusion to the race, but it was bad for Red Bull. That outcome would make Verstappen's task of catching and overtaking Hamilton much harder, as he would have to get past the lapped cars of (in the order that he would reach them) Sebastian Vettel, Charles Leclerc, Esteban Ocon, Fernando Alonso and Lando Norris before finally being able to chase Hamilton, who, in the meantime, would have been scampering down a clear track, even on old tyres. Practically an impossible task, which is why Masi came on the receiving end of some urgent lobbying from the Red Bull camp. First Christian Horner got on the radio:

> Horner: 'Why aren't we getting these lapped cars out of the way?'
> Masi: 'Christian, just give me a second. OK, my big one is to get this incident clear.'
> Horner: 'You only need one racing lap.'
> Wheatley: 'Obviously those lapped cars, you don't need to let them go right the way around and catch up with the back of the pack.'

This was a perceptive point by Wheatley. Because he knew the race director's job so well, the Red Bull man was pointing out to Masi that he had the option to release the lapped cars and restart the race before they had circulated all the way round and rejoined at the back of the field. It was normal practice to allow the lapped cars to do this, but it wasn't a requirement. Bear in mind that at this point, Masi had already decided (and put on the FIA messaging system) that lapped cars would not be allowed to overtake. Wheatley was suggesting a way that Masi could reverse that decision and still have time to resume the race.

> Masi: 'Understood. Just give us a second.'
> Wheatley: 'You just need to let them go, and then we've got a motor race on our hands.'

All of this was happening in a matter of seconds while Masi was also checking that the Latifi wreck had been safely removed, and that the track was clear. It was then that he did indeed reverse his earlier decision and wrote on the timing screens that the lapped cars of Norris, Alonso, Ocon, Leclerc and Vettel could overtake after all. The message appeared as their race numbers, so cars 4, 14, 31, 16 and 5. This was Masi's first unexpected deviation from the rules in that he made no mention of the other lapped cars in the pack, which, crucially, included Daniel Ricciardo and Lance Stroll, who were running between Verstappen and third-placed Carlos Sainz. The rules referred to 'any car that has been lapped' but there were three cars that had been lapped who weren't being allowed to unlap themselves. So while Verstappen was now being allowed a clear run at leader Hamilton, Sainz was not being given the same opportunity to tackle the Dutchman. It meant that Max didn't have to worry about defending his position, and could focus entirely on Lewis ahead.

Radio messages went out to Norris et al. to the effect that they should overtake Hamilton and the safety car. They did – while Stroll, Ricciardo and Schumacher asked their engineers why they weren't being allowed to do the same. Masi then put the message 'Safety Car In This Lap' on to the timing screens and the official messaging system. The safety car duly turned its lights off and prepared to drive into the pits. It was the end of lap 57, and the race was heading into the final lap.

The timing of this instruction to the safety-car driver was Masi's second deviation from the rules. The rule stated (and the established procedure was) that 'once the last lapped car has passed the leader the safety car will return to the pits at the end of the following lap'. This would have coincided with the last corner of the race, allowing Hamilton to drive across the finish line safely in front.

Instead, by ignoring that part of the rule, Masi had found a way to squeeze in a final racing lap. It was then that we heard Toto Wolff's urgent appeal to Masi: 'Michael, this isn't right. Michael, that is so not right!'

As the cars prepared for the restart, Verstappen closed up on Hamilton and pulled alongside. From the onboard camera view, he appeared to briefly poke his nose in front, causing Mercedes sporting director Ron Meadows to tell Masi that Max had passed Hamilton under the safety car, although in fact he had not fully passed him.

Verstappen was right with Hamilton as the green flag flew for that crucial final lap. Now it was a simple matter of physics – Lewis was helpless on the old, hard compound tyres he'd been on since lap 14. Max was on new soft tyres, with far more grip, and sure enough the Red Bull dived past at the hairpin. Lewis, who over the radio made his feelings clear – 'This is being manipulated, man' – tried to fight back, but Verstappen was comfortably ahead as the flag fell, winning the Grand Prix by 2.2 seconds and the world championship, with an advantage of eight points over his rival. It was then that Wolff voiced his utter frustration to Masi.

> Wolff: 'Michael, it was so not right. Michael, what was that? Michael? Race control?'
> Masi: 'Go ahead, Toto.'
> Wolff: 'You need to reinstate the lap before, that's not right!'
> Masi: 'It's called a motor race, OK?'
> Wolff: 'Sorry?'
> Masi: 'We went car racing.'

Masi had a right to defend his decisions, but his use of the phrase 'motor race' – the very words that Wheatley had put in his head as a suggestion just a couple of minutes earlier – did not go unnoticed by those listening across the world.

I watched all of this unfold from my position in the pit lane between the McLaren and Aston Martin garages. I had a big screen in front of me, and I was listening intently to Martin Brundle and David Croft in my headphones. The traditional Abu Dhabi celebratory fireworks went off as the flag fell, but I hardly noticed them as I tried to process what had just happened and decide what questions I should now be putting to the key people. As team members and VIP guests poured into the pit lane, I ran towards the Mercedes and Red Bull garages.

At Mercedes I found a team completely shocked, bewildered, confused. I think I used the word 'befuddled' at the time. There was no one at Mercedes who could understand what mechanism had just been used to circumvent the rules they felt the race should have been concluded under, namely that all the lapped cars should have been allowed to overtake the safety car and unlap themselves, and only then should the race have resumed on the following lap.

Attention turned to Article 48.12 of the FIA Sporting Regulations, which among other things noted that 'any cars that have been lapped by the leader will be required to pass the cars on the lead lap and the safety car.' As was to become apparent, that word 'any' was suddenly up for discussion – most people in F1 had always assumed that *all* cars had to be allowed past. However, there seemed to be little doubt over the other part of rule 48.12 that said, and it's worth repeating, 'once the last lapped car has passed the leader, the safety car will return to the pits at the end of the following lap.' That clearly hadn't happened.

Toto wasn't talking. I did a report into the commentary: 'They just don't understand what has happened at Mercedes, they're in a state of shock about what's going on.' I was trying to understand it as well, because we'd never seen the outcome of a race decided in this way, we'd only seen them decided according to how the rules were written. Meanwhile, at Red Bull, it was all kicking off. People were hugging each other and screaming, mechanics were dancing around the pit lane. One team had lost the 2021 world championship, and the other had won it. There were tears on both sides.

Mercedes immediately protested the result and then did something that seemed slightly strange: the team completely shut up shop. Wolff declined to be interviewed. Had Mercedes come out and been clear about which rules were not followed or had they even reminded Masi on the intercom during those last few laps of his obligations under rule 48.12, that would have been difficult for Masi to ignore and provided clearer context for everyone watching around the world.

Mercedes didn't do that. Emotions were running so high it was almost impossible to make cool-headed, detached public statements. Then there was a red herring the team pursued as part of the protest by suggesting that Verstappen had overtaken Hamilton under the safety car and he should be handed a five- or ten-second time penalty. All that did was divert everyone's attention off the main topic, which was that the race director had acted in such an unusual and unexpected way in his handling of the safety car. The issue of whether Max overtook Lewis under the safety car was fairly easily dismissed by the stewards – he drew alongside, but never had his car fully in front of Hamilton's.

The crucial part of that stewards' meeting was the discussion and the justification for Masi doing what he did with the safety-car rules. Mercedes was represented by Meadows, Andrew Shovlin and Paul Harris, a lawyer who the team had pre-emptively flown to Abu Dhabi just in case the championship had ended in controversy. Red Bull had Wheatley, Horner and Adrian Newey. According to people in the room, the Red Bull figure who was the most convincing and the most vociferous in defending their win wasn't Horner, or the rules man Jonathan Wheatley, but Newey. Frank Williams once called him 'the most competitive person in the pit lane' and here it was in action – Adrian Newey wasn't going to lose this fight.

After an initial exchange of opinions everyone went away to reflect and take the opportunity to gather more evidence. And it was Newey and Wheatley who found a hitherto little-known line in the FIA Sporting Regulations that transformed their case and could be used to get Masi off the hook for not adhering to rule 48.12.

Article 15.3 was, on the face of it, a simple confirmation of the split of responsibilities between the race director and the clerk of the course, the local official at each race who has responsibility for the marshals and track workers. It said that 'the race director shall have overriding authority in the following matters', and among the items listed was 'the use of the safety car'. It was this line that was highlighted by Red Bull, and was later relied on by the stewards to justify Masi's actions using the safety car as he saw fit.

This enabled the stewards to dismiss the Mercedes protest on the grounds that the race director has the overriding authority to control the use of the safety car, including its deployment and withdrawal. This was a curious decision based on a rule no one had paid much attention to before and certainly not relied on so heavily before, but there it was. In their judgement, the stewards acknowledged that 'although article 48.12 may not have been applied fully in relation to the safety car returning to the pits at the end of the following lap, once the message "safety car in this lap" has been displayed, it is mandatory to withdraw the safety car at the end of that lap', overriding the 'following lap' rule. It also agreed with Newey and Wheatley in ruling 'that Article 15.3 allows the race director to control the use of the safety car, which in our determination includes its deployment and withdrawal.'

Mercedes pointed out this was selective rule enforcing, that if the race had truly been run according to the rules, Lewis Hamilton would be world champion. Mercedes had called on the stewards to 'remediate the matter' by taking the positions of the cars at the end of the penultimate lap and determining that as the final race classification. This was an impossible ask. There was no mechanism in the rules to shorten a race retrospectively. The only time the classification from a previous lap can be taken is in the event of a red flag race stoppage that is not restarted, when the rules explicitly say that the result will be the race order at the end of the last lap before the red flag is shown (such as Brazil 2003, when Giancarlo Fisichella scored Jordan Grand Prix's last win).

With Mercedes's application to shorten the race by one lap rejected, the protest was dismissed, the race result made official and the 2021 world championship was decided. In the paddock we were all waiting for news. When it dropped, the atmosphere at Red Bull completely changed from one second to the other, as the team were finally able to celebrate Verstappen's championship win. Mercedes lodged a notice of an intention to appeal, which meant that in theory the race outcome was still provisional, but that didn't stop Red Bull enjoying the victory.

On TV, we congratulated Verstappen, wrapped up our programme (noting the appeal and how this would not be the end of the matter), and went off-air. It was a weird feeling on the bus to Dubai Airport that night. We were happy with the comprehensive coverage we put out that evening (the programme later won Sky Sports F1 a BAFTA award). However, in all honesty we weren't sure what had just gone on, or how Article 15.3 could be used to justify Masi's actions. I remember on the flight from Dubai to Heathrow someone asked me about what had happened. 'I've got no idea,' I said. 'You'll have to ask me in about a year's time.'

We didn't have to wait a year. The FIA gala prizegiving, the event at which Verstappen would receive his championship winner's trophy, was coming up the following week. It would be completely unprecedented for Max to be crowned 'provisional' world champion while a legal process was still unfolding in the background. With that in mind, just before the FIA event, Mercedes withdrew its appeal, saying in explanation that it didn't want to damage the image of F1 by rendering the championship provisional. Hamilton didn't attend the gala to collect his second-place trophy and was fined €50,000 by the FIA for missing it. Meanwhile, new FIA president Mohammed Ben Sulayem, recently elected to replace Jean Todt, promised a full investigation in the form of a special commission.

In a statement Mercedes noted that the team had originally 'appealed in the interest of sporting fairness, and we have since been in a constructive dialogue with the FIA and F1 to create clarity for the future, so that all competitors know the rules under which they are racing, and how they will be enforced. Thus, we welcome the decision by the FIA to install a commission to thoroughly analyse what happened in Abu Dhabi and to improve the robustness of rules, governance and decision making in F1.' Ahead of the full result of the investigation Ben Sulayem announced in February that Masi had lost his job as F1 race director and would be replaced for the 2022 season by Niels Wittich and Eduardo Freitas, who were both race directors in other FIA series. Masi parted company with the FIA, to return to Australia and the local touring car scene.

The FIA's full report finally came out in March 2022. Among the conclusions was an acknowledgement that in effect Masi had been doing his best, noting that he 'was acting in good faith and to the best of his knowledge given the difficult circumstances, particularly acknowledging the significant time constraints for decisions to be made and the immense pressure being applied by the teams.' However, the standout admission was that there had been 'human error' in the calculation over which lapped cars should be allowed to unlap themselves. Intriguingly it was Masi's deputy, American Scot Elkins, who had the job of telling the race director which cars should be waved through – the report suggests Elkins simply didn't realize that there were three more lapped cars. Meanwhile, the FIA said the rule about cars unlapping themselves would be changed from 'any cars' to 'all cars'. The FIA found their stewards had been correct in their judgement regarding which rule overrode the other and that the famous rule 15.3 legitimized Masi's use of the safety car in the way he did.

And that was it. Mistakes had been made, but Masi, it was felt, had acted in good faith and within his authority. This begged the question, though, if Masi hadn't done anything wrong in the FIA's summation, why was he removed from his job?

I have absolutely no problem with anything that Max Verstappen did in the car at the end of that race. It was the easiest strategic decision of the season for Red Bull to pit him under the safety car, and he drove perfectly to get the best advantage from the new soft tyres, performing a clean overtake on Hamilton, and winning the race to become a deserving world champion. On that night, he successfully played the hand that was dealt to him. Of course, had Latifi not crashed, Hamilton would have matched Verstappen's nine wins, and finished eight points clear to clinch his eighth world title. It was because of Masi and his officials' unusual actions that Lewis didn't win the race, so clearly, Hamilton would also have been a deserving world champion.

I also understand Horner and Wheatley (and later Newey) trying their luck with the referee. Granted, it's not the same as appealing for a penalty in football or an LBW decision in cricket, because in those instances, you're not encouraging the referee to enforce some rules of the game but not others – but the result is the same. Many people inside and outside the F1 paddock accused Wheatley and Horner of being unethical or unsporting in trying to influence Masi to improperly execute the rules. The Red Bull duo deny that and I can see their side of the argument – they had to try something. It was their job and their responsibility to their driver and team to try anything they could to help Max win. Whether Horner and Wheatley ever – for one single second – believed Masi was going to agree with them and carry out their suggestions is a different matter. The looks of surprise on their faces immediately after the race suggested to me that they never in a million years expected Masi to go along with their appeal of 'you only need one racing lap'.

Toto Wolff's belief was that Masi had been conditioned to accept Wheatley's suggestion because Wheatley had helped him out so many times previously – Wolff even described Masi and Wheatley's relationship as a 'bromance'.

As for Masi, for all his faults I don't believe (and there is no evidence that) he had a preference either way in terms of Red Bull or Mercedes, Verstappen or Hamilton, that he wanted one driver to be world champion over another. The FIA investigation concluded he was acting in good faith, and doing what he thought was the best thing to do in the situation. That is not to say that he made the right calls that day.

Wittich and Freitas took turns as race director in 2022, and this time the FIA was careful to keep both men well away from the media spotlight. There was also an abrupt end to the broadcasting of any pit wall intercom communications with race control – and team principals were banned from voicing their opinions. Henceforth only the sporting directors would be allowed to talk to the race director and then only to ask questions, not to suggest actions. Freitas soon departed, and Wittich was handed the race director job full-time. However, in November 2024, Wittich was himself sacked by Mohammed Ben Sulayem – a development that Masi may have watched from afar with interest.

Max Verstappen won every championship from 2021 until 2024. Jonathan Wheatley left Red Bull to become team principal at Sauber/Audi. Christian Horner never left the spotlight and was subsequently sacked by Red Bull in July 2025.

As for Lewis Hamilton, he had been set to become the undisputed, greatest F1 driver of all time, beating Michael Schumacher to eight world titles. In the weeks and months after Abu Dhabi, he seriously considered quitting F1. With the support and advice of his family and friends, Lewis decided to return for 2022, not wanting the last word on his career to be what happened in Abu Dhabi. He wouldn't win again until the 2024 British GP, when the emotions came flooding back. After that race, Hamilton said, 'Honestly, when I came back in 2022, I thought that I was over it. And I know I wasn't, and it's taken a long time to heal that kind of feeling. And that's only natural for anyone that has that experience. I've just been continuing to try and work on myself and find that inner peace day-by-day.'

F1 has never been more popular, so it's difficult to judge to what extent the events of that night in Abu Dhabi damaged the image and integrity of the sport. Just like all the other great stories that make up F1's fascinating history, though, we shouldn't be afraid to talk about it.

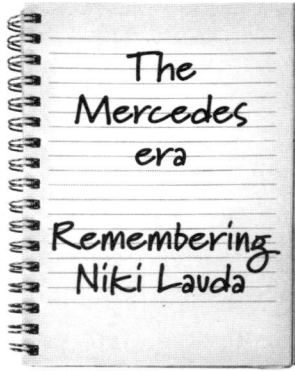

The
Mercedes
era

Remembering
Niki Lauda

Chapter 21

The Eight-Pointed Stars

———

Pressure is a constant in F1. It's effective in terms of driving performance, but it's not healthy. So when Toto Wolff finally activated his personal pressure-relief valve in an Abu Dhabi nightclub just a few hours after that fateful 2021 Grand Prix, he really went for it. Lewis Hamilton might have lost the drivers' championship, but his 18 points for second place won Mercedes their eighth consecutive constructors' title.

As 'Freed From Desire' by Gala blasted out of the speakers, Toto crowd-surfed off the DJ decks into the waiting arms of his Mercedes mechanics, crisp blue shirt getting lost in the throng. There followed a group rendition of Queen's 'We Are the Champions', again. But there was something about the way Wolff threw himself into the celebrations that suggested a man more interested in burying his feelings of shock and outrage over the way the last race had unfolded, than savouring the actual championship they had won. He had clearly decided the only way to cope that night was to let go and blow off some steam. However, as a leader he had just achieved something remarkable. Hamilton might have lost the drivers' title at the very last gasp, but nine years after writing that report for the Mercedes board about why their F1 team wasn't winning, Toto Wolff had led it to an eighth world championship. It was a phenomenal achievement. No team had ever won eight in a row – the closest anyone came was Ferrari's six titles from 1999 to 2004.

There were many stars in that team, both at the track and back at their Brackley factory. Most are still there, which makes their failure to master the ground-effect rule which lasted from 2022 to 2025 even more baffling. Lewis Hamilton won six of his seven titles with Mercedes and held out for another winning car for as long as he could, before electing to leave at the end of 2024 for a longer term deal at Ferrari. It was a move that sent shockwaves through the sport when it was announced, but it just felt right that the most successful driver of all time should drive for the most famous team of all time. Time will tell if he can win that eighth world title in red, but when I look back on the story of Sir Lewis Hamilton's F1 career – and I've reported on every season of it – a picture emerges of at least seven key characteristics that have kept him at the top.

1. Natural speed. When Lewis was struggling to match the single-lap qualifying pace of his teammate George Russell in 2024 he said, 'I'm definitely not fast anymore.' He would then deliver the kind of lap times in the races the next day that proved that he was definitely still fast. There are neurological and physiological reasons why athletes in their 40s are not as performant as those in their 20s. There's physical strength, balance, reaction speeds – all

are applicable to racing drivers. The time it takes for the eye to see, the brain to process and the body to react gets greater with age. These are well understood in such sports as downhill skiing, which is similar to motor racing in terms of reaction times and physical movements. We're talking tiny differences here, tenths of a second at most, and there are so many steering-wheel and throttle and brake inputs as to make the effect of age negligible. While Hamilton continues to try and rediscover his qualifying edge, Fernando Alonso, three years his senior, is proof that drivers don't lose their basic speed. And given that Hamilton's (and Alonso's, for that matter) natural speed started at a higher level than most in F1, any age-related loss is, in the races at least, yet to become a factor.

2. Thirst for success. Having reported on all of Lewis's championship seasons, what has consistently impressed me has been his relentless drive to achieve more. He's as competitive a personality as any other driver, but there's something about him that treats the beginning of each season, with 2025 at Ferrari being the best example, like it's his first. He has an ability to reset himself over the winter and always hit the first race with maximum motivation. Sure, we've occasionally seen him wanting to park his car in a hopeless race, only to be dissuaded by his team, but that intention was always to 'save the car' for better performance in a future Grand Prix. In the end, he never gives up if there's a chance of a win or points for the world championship.

3. Intensity of focus on a race weekend. Maybe it's because Hamilton is F1's only current worldwide household name (just as Michael Schumacher and Ayrton Senna were before him) but

there's always more attention surrounding him at a race track than anyone else. A bigger crowd in the paddock, more fans lining up at the gate, always more of a fuss. It would be a huge drain on anyone's mental energy to absorb and react to that constantly, so with the exception of the drivers' parade, which Hamilton uses as a pre-race pump-up, he has an ability to shut out the rest of the world and remain laser-focused on his driving.

4. Consistency. This is by no means unique, but no driver has more experience than Hamilton at putting a world championship together in terms of scoring the points week in, week out. In a title fight there are few drivers better at maximizing their points and being consistent over 24 rounds. Even when he can't win, he's racking up the points – his seven world titles being the obvious result.

5. Work ethic. Lewis is honest with himself and habitually looks for areas that require improvement and then puts in the work to fix the deficiencies. Even at the beginning of his career the pace at which he would identify areas where he was weak and then improve them was particularly impressive.

6. Driving style. In this period of F1 when the races are predominately decided on who can get their tyres working best and lasting the longest, Hamilton has been adept at driving around the peculiar characteristics of the Pirellis. It's not a particularly sexy subject, but being a 'tyre whisperer' is something he can turn his skill set to when required.

7. Race craft. Lewis's father Anthony once said to me, 'The thing people don't fully appreciate about Lewis is that he's a mercenary. He's relentless, ruthless. A cold-blooded, merciless racer.' Again, not unique, but experience has given Hamilton such a massive databank of knowledge, of corner profiles, grip levels, hybrid power unit electricity generating zones – anything that can help predict who will be vulnerable to an overtake where and when. Plus, his technique, being clean, clinical and contact-free when he gets there; he remains one of the most effective racers and overtakers on the grid.

And to add to all of these, my own experience of Lewis has me agreeing with his ex-McLaren head of communications Matt Bishop, who called him 'a first-class human being'. It would be impossible not to admire the work Hamilton does when he's not at the track, using the power of his platform and status to effect positive change in the world. He has a heartfelt desire to help people less fortunate or in less advantaged positions achieve as much as possible in their lives. He established Mission 44, a charity that works to build 'a fairer, more inclusive future for young people around the world and to overcome social injustice'. I've seen kids and young people from every continent visit the F1 paddock on many occasions with Mission 44 and be inspired by what they see. It's already having an incredible effect.

Back in 2020 Lewis also put his own money into establishing The Hamilton Commission, which produced a report in 2021 titled Accelerating Change: Improving Representation of Black People in UK Motorsport. The report explores the barriers to recruitment and progression, which start in early life and last throughout young people's educational journeys, and provided 10 recommendations to address them. In 2024, the F1 organization, the FIA and all the teams adopted the recommendations of The Hamilton Commission in the form of a Diversity and Inclusion charter, the first of its kind in F1, and something that promises to get previously under-represented groups into motor sport. Nobody else – driver, team principal, team owner, administrator or governing body – has done as much as Lewis Hamilton to effect positive change in F1.

The 2021 controversy aside, the other title that got away from Hamilton was in 2016, when Nico Rosberg beat him to the world championship by five points, despite Lewis winning one more race. Reliability issues such as Hamilton's engine failure in Malaysia paired with Rosberg's consistency and an excellent drive under pressure at the deciding race in Abu Dhabi sealed the title, making Keke and Nico Rosberg the second father-son duo to win the F1 world championship after Graham and Damon Hill.

Nico is now a colleague on Sky F1, but we've known each other since his GP2 win in 2005 and his arrival in F1 at Williams in 2006. I'd often ask him questions about all sorts of topics, from Spy-gate to Crash-gate, largely because he could always be relied upon for a good answer. Nothing has changed in that regard, and as an analyst you can put any question his way and not only will you get an answer, but Nico will usually throw a question back, so take note that you have to be ready for it.

His approach to beating Lewis Hamilton in 2016 was forensic. He'd learned everything about Lewis that he didn't know already over the previous three seasons, including how to use the media in an attempt to get under his skin. It's hard to tell if that ever worked, but Rosberg's dedication and commitment over that 2016 season were so intense as to be mentally unsustainable and certainly incompatible with an ordinary family life. His subsequent decision to retire shortly after winning his world championship seemed a fairly easy one.

These days Rosberg is much more relaxed. I suppose that having achieved your life's ambition and retired at the age of 31, financially secure for the rest of your days, means you can afford to be easy-going. More recently he has used everything he learned about himself and the power of human relationships to become a very successful businessman and investor.

One evening following the 2017 Japanese GP at Suzuka, Nico and I found ourselves in the back of a Mitsubishi Outlander for the two-hour trip from Suzuka to Osaka's Kansai Airport. I would have happily dozed off in the back of the car, or sat and listened to some music. However, typically for a racing driver, with brains used to processing information at speed, Rosberg gets bored quickly. So we soon got into conversation about a whole variety of subjects, from TV ratings to F1 teams, family to favourite airlines and flight routes. What caught Nico's attention was my answer to one particular question: 'Who's your favourite driver? Who do you really like?'

'Apart from Senna?' I replied. 'Erm . . . Alexander Wurz,' I said, namechecking the lanky Austrian who drove for Benetton, McLaren and Williams. 'Really?' said Nico, surprised. 'Why?' 'Well, I suppose because he was the only driver in my time working in F1 that I really identified with. We're the same age, and while he was very thoughtful about race strategy, setup and tactics during his driving career, when I interviewed him, he often surprised me with his breadth of intelligence and knowledge about other things in life. He's a great guy.'

'Oh,' said Nico. And then, after a pause, 'What about current drivers?' I explained that there were things I liked and admired about all of them, but there wasn't one who was my favourite. 'I honestly don't have a preference who wins and who doesn't.'

'Yes, but . . .' he countered. 'Personally, you must have a view on who the current drivers are as people?' We discussed how Hamilton is an awesome driver but can be distant while he's lost in the focus of a Grand Prix weekend. How Fernando Alonso continues to impress and amuse, how Nico Hulkenberg is smart and quick, and how Daniel Ricciardo is generally lovely and adored by everyone he meets.

As the elevated motorway passed south of the city of Nara, I asked Rosberg how much of what we see of the drivers is their true character? They're heroes to many because they have skills the rest of us don't, and are expected to perform to their maximum potential, under intense pressure and scrutiny, every other Sunday afternoon. But do we ever really know them?

'Ah', said Nico. 'Definitely not. It's hard. People don't realize how difficult it is to be perfect, to be on top of your game, week after week. And how if you make one little mistake, whether it's a crash or a spin, not qualifying on pole, or a race where you're even the slightest bit off-form, everyone asks why you're suddenly useless.'

At that point the Mitsubishi fell silent as we both stared out the window and I tried to remember how many times I'd asked him something like that.

It's impossible to truly judge F1 drivers on anything but the most superficial level as we're asking for perfection every day of their working lives. Applying that standard to us more ordinary folk proves the point. We might go to work and do a decent job and come home feeling like we achieved something, but the reality is that most of us don't hit perfection every day. I certainly don't. However, that is what's demanded of F1 drivers, and when they don't achieve perfection, it gets noticed.

I tried to make the same point to Rosberg's Mercedes successor Valtteri Bottas at the end of a weekend spent filming with him in his hometown of Nastola in Finland. We were talking by the side of his local kart track, where he'd learned and honed his driving skills, in dry, wet and even snowy conditions. I thought it reasonable to observe that, just like Rosberg had found, Bottas might have to accept that on some days, Hamilton is simply unbeatable in the same car. And that's the way it is. 'No, I don't think that,' Valtteri said, his blue eyes narrowing against the cold. 'If I thought that, even for a second, I would just not show up in F1.'

I shouldn't have been surprised. Bottas put up a very decent fight against Hamilton over the five seasons they were teammates in 2017–21, winning ten races and millions of fans worldwide. He's one of the drivers of recent times who reminds me most of the seventies and the James Hunt era. That weekend in Nastola was one of my favourite road trips – enjoying some serious sauna time, before lowering ourselves slowly into the adjacent icy lake. Quite a buzz.

Bottas was a huge part of Mercedes winning those eight constructors' championships, and after a spell at Sauber he returned to the team as third driver. While he found many familiar faces, some of the key people with whom he previously worked are no longer in the Mercedes camp. As the years passed, so the dream team unravelled, as F1 dream teams always tend to. First to step back was Andy Cowell, the head of the Mercedes engine division. After a spell on the sidelines, he became team principal at Aston Martin and is building his own dream team with Adrian Newey as technical partner. James Vowles is trying to do the same at Williams, his former role at Mercedes having developed far beyond that of head of strategy, before he left to become team principal of the Grove outfit at the start of the 2023 season.

Aldo Costa was a key engineer who had spent many years at Ferrari, and was known at Mercedes as 'Mr Suspension', thanks to having extensive experience of vehicle dynamics and getting the mechanical platform of a car to work smoothly with its aerodynamics. Costa's return to Italy and a new role at chassis builder Dallara was a key loss, as was Loic Serra, who worked on chassis design and general tyre dynamics. He joined Ferrari as its chassis technical director in 2024, and is now working once more with Hamilton. Also gone is Mike Elliott, a gifted aerodynamicist who was promoted to Mercedes technical director and then moved sideways to chief technical officer, neither of which really worked out.

Despite these departures, many key players remain in the Brackley camp, some of whom go back as far as the BAR, Honda and Brawn GP days. Sporting director Ron Meadows keeps the race team going and stays on top of the rules, while Andrew Shovlin heads up trackside engineering with Peter Bonnington. And then there's technical director James Allison. A self-confessed armed forces brat (his father being former Air Chief Marshal Sir John Allison), James is the kind of person who ought to go down in history as one of Britain's greatest engineers. He's the kind of person who, had he been alive in either world war, would have come up with a way of shortening it considerably, and who, if you worked for him, you'd be willing to jump off a cliff for, knowing that he would have invented some clever net device to catch you and that he'd be jumping too. Were he not colour-blind and too tall, he would have made an excellent Spitfire pilot. He scratches that itch flying vintage and not-so-vintage aircraft around the UK for fun.

I first met James when he was at Ferrari, leading the trackside aerodynamics team under Ross Brawn. I wondered who this guy was who spoke good Italian with a very British accent. It has been a great pleasure to know him ever since. If anyone can lead Mercedes out of the doldrums and back to brisk, blue skies it's Allison.

He has also been a valued sounding board for Toto Wolff in recent years, in much the same way that Niki Lauda used to be. However, the three-times world champion has proved to be completely irreplaceable in the Mercedes F1 camp, so much so that his trademark red cap still sits on top of his radio and headset on the rack inside their garage.

I used to interview Niki after most races, and like anyone who knew him, developed an immense respect for his many achievements. Having given the Grim Reaper the slip in 1976, death finally caught up with Lauda in 2019, when he succumbed to an illness related to his second set of kidneys. He was on his second set of lungs, too, following a transplant in August 2018. 'Whatever it takes to make things right.' This was Lauda's life philosophy, a time-efficient route to satisfaction – be it major organ transplant or perfecting race car setups, it made no difference. Something isn't right? Fix it. Quickly!

As a driver Niki was very good, but he perhaps wasn't naturally gifted with raw speed in the style of Jim Clark, Ayrton Senna, Michael Schumacher or Lewis Hamilton. Rather, his strengths were in understanding his car's problems and knowing, with canny accuracy, how to resolve them. These days, Formula 1 engineers need banks of computers fed by thousands of sensors to tell them what the car is doing. With Niki, the computer was in his head.

The human computer worked. Having bluffed and bought his way into F1, Lauda won the 1975 and 1977 world championships for Ferrari, reaping what he'd sown in perfecting his cars. He would likely have won in 1976, too, were it not for the fateful accident at the Nürburgring, the resulting fire from which burned his face and choked him with noxious smoke. Doctors at the hospital didn't rate Lauda's chances of survival, and called a local priest to administer the last rites. But Niki wasn't ready to die that day. His injuries were just a problem he needed to fix.

It was the same for everything else in Lauda's life: from cars to airlines to wives. Married twice, Niki fathered five children and created three airlines. Lauda Air, which provided a focus when he first hung up his racing boots in 1979, was commercially successful (it was later bought by Austrian Airlines), but suffered a tragic crash in 1991 which killed 223 passengers and crew when a thrust reverser on a Boeing 767 deployed in flight. Lauda identified the cause with Boeing and made the American company fix it, ensuring a similar technical failure could never happen again.

He joked that he was running out of names after branding subsequent airlines Fly Niki and Lauda Motion. The discipline of flying appealed to the computer in his brain. An accomplished pilot, he flew every aircraft his companies operated, from huge

Boeing 777s to his private Bombardier Global 7000. But motorsport was never far away. Once Lauda Air had been sold, Niki received an offer from Ford to help run the Jaguar Racing team. His direct style worked wonders with drivers like Eddie Irvine, but didn't go down well at Ford's Dearborn HQ, and his bosses dispensed with the Austrian's services. He took a job as expert pundit with German broadcaster RTL where he originated the role of paddock insider, complete with forthright opinions, long before Eddie Jordan picked up a microphone. Niki frequently upset the thin-skinned, but he never lost anybody's respect.

The last race Niki Lauda attended was the British Grand Prix at Silverstone in June 2018, not long before his life-saving lung transplant. Ironically Mercedes was beaten that day by Ferrari and Sebastian Vettel, with whom he shared a natural affinity – the dirty jokes, the no-bullshit attitude, the stubborn refusal to accept an unsatisfactory situation. That was Lauda's way, not because he was a difficult person, but because his history had taught him to speak freely without fear of the consequences, aware there was a good chance the next day might be his last. We'll never see his like again, but every time I walk through the Mercedes garage and past his red cap on the radio rack, I hear his voice, in that thick Austrian accent: 'Hey, it's you again! Piss off back to the pit lane!'

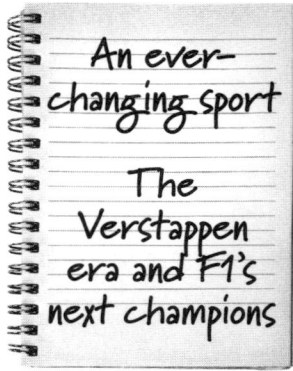

An ever-
changing sport

The
Verstappen
era and F1's
next champions

Chapter 22

The Finish Line

———

It was a beautiful March morning as race week began ahead of the 2000 Australian Grand Prix. Michael Schumacher was about to kick off what would become the first of his five straight world championships with Ferrari, and we had landed at Melbourne's Tullamarine Airport to start ITV Sport's fourth F1 season.

While James Allen, Louise Goodman and I stared out of the taxi-minibus window at the buildings that had sprung up since our last visit, a familiar voice from the front passenger seat burst the silence. 'You know the thing about F1? It has a unique ability to reinvent itself. Year after year, it finds a way to be new and interesting. Don't know how. Marvellous, really.' And with that, Murray Walker turned back to the windscreen and his thoughts, fingers drumming on his right knee.

Exhausted from the flight, I didn't think much of Murray's comment at the time. However, as the years have gone by, I've realized what an astute observation it was. Whether because of drivers moving teams, new drivers or new teams coming in, new race venues added to the schedule, or rule changes that threaten to disrupt the competitive order, F1 is always moving forward, always changing, creating new excitement every season that few other sports can match.

My grandfather, as he advanced towards the grand old age of 103, had an observation about the secret to a long and happy life: 'You always have to have something to look forward to.' Murray knew this better than anyone. At the first race you don't yet know what that new thing will be, or the story of the championship, but no question, you're looking forward to finding out.

One of my favourite things about the start of each new season is living through events that you will later come to recognize as significant. One of the big talking points of late has been whether Max Verstappen can add to his title tally in this, the Verstappen era. An era that saw its roots planted on Saturday 19 July 2014, qualifying day for that year's German Grand Prix.

At the time Max was only 16, but many people in the F1 world already knew how successful Jos Verstappen's boy had been in karting, how quickly he had adapted to F3 machinery, then blipped on the radar of the F1 teams. Hockenheim was the first time that Max had been in the F1 paddock, not as the young son of a driver, but as a driver in his own right, looking to break into the top level. It was the first time that he was there purely on business with the intention of walking away with a contract that would change the rest of his life.

Jos and Max had a meeting with Toto Wolff and Niki Lauda at Mercedes, but they couldn't offer him a race drive while Lewis Hamilton and Nico Rosberg were fighting for the world championship, and wouldn't be able to place him at a customer team for a year or so. In contrast, Red Bull had junior team Toro Rosso for just this purpose, and Helmut Marko and Christian Horner were only too ready to turf out an incumbent driver as soon as a potential megastar came along.

I spotted Jos and Max as they were leaving the back exit of the paddock. I'd known Jos since my first season in 1997. Indeed, my debut F1 television feature had been about Jos, following a great race in the wet 2001 Malaysian GP, when he drove brilliantly to finish seventh from 18th on the grid in an otherwise uncompetitive Arrows.

I greeted them and asked how things had been going. It was just a quiet little visit, Jos said, catching up with people and showing Max around. I introduced myself to Max, wished him good luck, and said that I was sure that he would make it into F1 quite soon. Jos, playing his cards close to his chest, said there was a long way to go before F1 was an option, that nothing had been decided. However, that day in Hockenheim had been very productive, and Max was announced as a Red Bull stable driver a month later. Having turned 17 he then drove in a practice session at that October's Japanese GP for Toro Rosso as part of the preparations for his debut season in 2015.

Jos 'The Boss' Verstappen's was a career that never quite fulfilled the promise it showed when he arrived in F1 in 1994 touted as the next big thing. He was viewed as a potential world champion, an image boosted by the fact that he drove for Willi Weber's F3 team three years after Michael Schumacher. He was signed as a Benetton reserve driver by Flavio Briatore, and soon graduated to a race seat as a substitute for JJ Lehto after the Finn fractured a vertebra in a testing accident.

Driving alongside Schumacher, Jos had a couple of third places, but he was most famous for being engulfed in a fireball when a botched pit stop sent fuel spraying all over his car at the German GP. Like many TV viewers, I had been impressed by his insouciant wafting away of the fuel fumes at the time of the initial spill, only for the whole car to be set alight, forcing the Benetton pit crew to dive for cover. Jos suffered some minor burns, but was otherwise unharmed.

Spells at Arrows, Simtek, Tyrrell and Minardi followed, but Verstappen never got another chance at driving a potentially race-winning car. His hindsight has been Max's foresight – Jos learned the hard way about bad career moves, understood only too well the frustration of driving poor machinery and always looked for the quickest cars for his son.

Max's career gained even more momentum when he broke Sebastian Vettel's record as F1's youngest race winner, triumphing on his first outing with the senior Red Bull Racing team at the 2016 Spanish GP. He was only 18 at the time, but he has always looked grown up, unlike other young F1 drivers who seem to mature before your eyes, Lando Norris being a good example.

What has developed over his decade at the top level has been Verstappen's assured personality. He's focused on the few things that really matter to him in life – essentially family and racing. It's a refreshing character trait. He has no time for any of the off-track distractions that so often complicate life for F1 drivers.

The Verstappen era has been interesting, as the way he has won each title has varied so greatly. The 2021 season ended as previously described, and it came at the end of a gruelling year full of drama, forging a mental toughness that has since served Max well.

The 2022 season saw the beginning of the ground-effect rules, which Red Bull's designer Adrian Newey understood better than anyone else. Max took advantage of the strong car and secured his second title with ease, winning 15 of the 22 races. However, that was a modest tally compared with 2023, when the RB19 was so dominant that Verstappen logged 19 victories, with teammate Sergio Pérez scoring two wins and only one escaping Red Bull, instead going to Ferrari's Carlos Sainz. It was a level of dominance that had not been seen since Michael Schumacher's Ferrari era. Then in 2024 Max got off to a flying start before McLaren and others caught up with Red Bull's pace, and thereafter he fought a rearguard action, still managing to secure his fourth title.

In a sense his approach to driving is akin to Schumacher's. Max doesn't give presents and he doesn't back down. He's incredibly skilful in wet weather conditions and has a fearless attitude in wheel-to-wheel battles, often going for an overtake at the first opportunity when his opponents realize that 'it's Verstappen' behind. He also has the Schumacher-style edge that can take his racing into the grey areas of fairness. Particularly evident in his on-track battles with Lando Norris, Charles Leclerc and Lewis Hamilton, Max knows the driving guidelines better than most, and is adept at putting them into practice in ways that are hard to penalize.

At the beginning of the 2025 season, Max was only 27 years old, an age that gives him the luxury of having enough time to achieve everything he wants to in F1, and then still have time to go and race in other categories. He's made no secret of his desire to try his hand at endurance races such as the Le Mans 24 Hours and the other major 24-hour races at Daytona and at the Nürburgring.

Essentially, the more time he can spend driving and racing, and the less time doing all the other tiresome tasks an F1 driver has to commit to, the better.

In the meantime, he has Grands Prix to contest, and rivalries to revisit. The very public spat with George Russell towards the end of 2024 was a good character study of both men. In the closing stages of qualifying for the Qatar GP, with both drivers on their build-up laps before starting proper qualifying laps, Russell accelerated through a couple of corners to find Verstappen on the racing line. George swerved to avoid an accident, and reported to his team on the radio what he saw as Verstappen's 'super dangerous' driving. As both were on out-laps, nothing much was thought of it, because no harm had been done to either driver's competitive result.

However, after the session the incident was investigated, and both Max and George were called to the FIA stewards to explain themselves. According to Verstappen, Russell tried his hardest to get Max a penalty, and succeeded – Max was duly given a one-place grid drop for 'driving unnecessarily slowly' on the racing line. There were two viewpoints. Russell claimed that Max was trying to slow his preparation lap down, which would cool the Mercedes W15's tyres, and compromise his lap. Verstappen insisted that he was courteously staying out of the way and preparing his tyres for his own lap. Max later said of George that he'd 'never seen someone try to screw someone over that hard,' and that he'd 'lost all respect' for him.

Verstappen and his team were furious with what they saw as an unjust penalty. Max was so determined to take the lead and seek justice for his grid drop that, coming out of the Saturday evening hearing, Russell said that Verstappen threatened to 'put him on his fucking head in the wall'. Verstappen denied he 'said it like that'. The next day, according to people who were present at the assembly point for the drivers' parade, Verstappen, still angry about the grid penalty, greeted Russell with 'I hope you're happy with yourselves, you and your FIA buddies.'

Compared to qualifying, the race was tame. As if to prove that an angry Max is a quick Max, or maybe he was just fired up because of his penalty, Verstappen took the lead from Russell at the start and never looked back, winning the Grand Prix by six seconds to Charles Leclerc. Russell was fourth.

One aspect of George Russell's character that many people miss is that, despite the gentlemanly exterior, he's a tough customer, completely comfortable fighting his corner when he feels he's in the right. For example, a week after Qatar, when Russell arrived at the Abu Dhabi track, he was in no mood to let the Verstappen situation lie. I'd spent the day interviewing drivers at the TV pen but hadn't yet heard any earth-shattering stories.

A minute before his time slot, up strode George. He called me over to the edge of the pen: 'Make sure you ask me a follow-up question.' 'What do you mean?' I asked. 'I'm going to give it back to Max, I've had enough of him bad-mouthing me in the press and I'm going to call him out on his bullying tactics. I know you're only supposed to ask one question, but never mind that, I'm up for as many questions as you like.' Before checking to see if he was wearing a *Drive to Survive* microphone, I asked him if he was absolutely sure he wanted to escalate what was essentially last week's story and start a new fight with Max, never an easy battle to win. He said he was positive, and away he went.

'I've known Max for 12 years,' George said on camera. 'I've respected him all of this time, but now I've lost respect for him because we're all fighting on track, and it's never personal. Now he's made it personal, and someone needs to stand up to a bully like this. So far, people let him get away with murder.' Max called George 'a loser' and on it went – an old-fashioned grudge match that shows no sign of cooling down.

As for the other potential future world champions, Lando Norris, Oscar Piastri, Charles Leclerc and Carlos Sainz have all demonstrated that they know how to piece together winning race weekends. All could be champions if they get a chance in the right car, and certainly as of 2025 the McLaren duo of Piastri and Norris have the best package at their disposal. I'd add Alex Albon to that list, even though, at the time of writing, he's yet to win a Grand Prix, while Kimi Antonelli has the might of Toto Wolff and Mercedes behind him, and, at the start of the 2025 season, was the best-prepared rookie I've seen in years.

In many ways racing is the easy part of the job for the current generation of F1 drivers. Physically, they're top of the scale, able to withstand prolonged G-forces and attack from every dimension while trying to drive at speed and control a car that would happily kill them if not handled with care. Even though this is a sport practiced sitting down, they are incredible athletes. But they also have to balance driving with the fame and attention that comes their way, which can flick from acclaim to abuse in seconds in an era dominated by social media. As a result, they have developed different ways of coping with the pressure. Lando Norris has been particularly open about his mental health in a way that would have seemed inconceivable to the drivers of Senna and Schumacher's generation. The money and the lifestyle might be one thing, but it's also evident that those things come at a cost.

It's hard to think of another sport that makes such demands of its participants. Although F1 is unquestionably a business, there are enough moments of pure sporting drama, passion, controversy and achievement taking place throughout the field to keep even the most cynical rev-head satisfied. The more you know about the personalities of F1 and the engineering technology of the cars, the more you want to know. That's sometimes hard, as it's a secretive sport, but the paddock thrives on gossip. Teams and the media arrive at the circuit early to beat the traffic, they leave late, and as there are only a few hours of track action each day, there's a lot of time for everyone to hang around and talk, about drivers, cars, teams, people and internal politics. Information is currency, and scraps are keenly traded. It's important not to lose perspective and not to forget that it's all just part of the show in the giant circus that is F1. A circus that pitches its tent, sprinkles its magic dust and performs, before packing up and moving on to the next town.

Which leads us back to the finish line, and what Anthony Davidson called my 'made-up job'. I hope you have a sense of what it entails and how addictive an environment it is, and the amazing stories I've witnessed. F1 has changed in so many ways over the last 30 years – more professional, more serious, more commercial and certainly more valuable. But it's also still fun, provided you don't take it too seriously.

F1 means as much to me now as it did back when I was in my parents' kitchen scouring the papers for snippets of information or the results tables. I'm still fascinated by the cars, interested in the drivers, enjoy the characters, try to keep on top of the ever-changing tech, and keep on the good side of the team bosses – not to mention the sounds, the smells, the buzz, the roar of the crowd after an amazing overtake, the circuits and countries I get to visit.

If you've recently started following F1, welcome to the club. It makes me happy to think of new people coming in and discovering the sport I've loved so much for so long. I hope you've enjoyed reading some of the stories of the last few decades – the backstories, if you will. I've always been fascinated by the details, and when you put them together it makes those stories come to life. To those readers who've been as obsessed as I have, I hope you've enjoyed this trip through F1 eras past, from Schumacher and Alonso to Hamilton and Verstappen. The sport has evolved, but the essence remains the same. A unique combination of groups of people coming together to create extraordinary machines, raced by singularly talented individuals, with so many variables you never know what's going to happen in the next moment, much less across a season. The only thing I can be sure of is that I won't have long to wait before the next amazing story.

Thanks for reading, and I'll see you next time you switch on your television, tablet or phone and watch a race with us on a Sunday afternoon. I'll do my utmost to bring the pit lane into your front room, connect you with the drivers, take you behind the scenes and deliver as much access, news, gossip, information and entertainment from the paddock as possible. Come to think of it, everything you need to be an F1 insider.

Index

Abu Dhabi GP: 2010: 232–4; 2021: 287–8, 292–308

accidents and fatalities 96–9, 100–1; controversial 103–4, 107–8; helicopters 9–10; 'planned' 42; Renault 'crash-gate' investigation 176–9

air travel 117–20

Albon, Alex 330

Allen, James 31, 34, 35, 46–7, 61–2; advice to TK 105–6; successor to Murray Walker 70, 79

Allison, James 318–19

Alonso, Fernando 42, 89–90, 114–15, 116, 311; 2008 season 172–3, 175–7, 182; Abu Dhabi GP 2010: 232; and Aston Martin 246; and Ferrari 240, 242; and Hamilton 143, 147–50, 152–7, 159; leaves F1 temporarily 245; and McLaren 143, 162, 244; and Renault 172–3, 175, 245–6; and Schumacher 103–4, 107

Arden (team) 215–16

Arrivabene, Maurizio 242, 243, 247

Arrows (team) 29–30

Australian GP: 1994: 55–6; 1997: 36–8; 2020: 274–81; 2022: 168

Austrian GP 2002: 84–91

Azerbaijan GP 2021: 290–1

Bahrain GP: 2020: 100–1, 266

Barrichello, Rubens 15, 191, 193, 196; and Schumacher 82–9; and Williams team 200

Barwick, Brian 69, 80–1

Bauer, Jo 263–4

BBC 31, 130, 184–5, 193; F1 Forum 193; From the Pit Lane 255–6

Belgian GP 56; 2000: 65–7; 2024: 263–4

Bell, Bob 179–80

Ben Sulayem, Mohammed 304, 305

Bianchi, Jules 96–8

Bishop, Matt 272, 313

BMW Sauber team 193, 227–30, 248

Bond Muir, Catherine 270, 272, 273, 274
Bottas, Valtteri 220, 317
Brawn, Ross 14–15, 85–6, 87–8, 90–1, 278; Brawn GP 187–200, 206; double diffuser issue 190–4; and Honda 15, 185–9; and Mercedes 221–2; and Schumacher 108, 111; and Toto Wolff 222
Brazilian GP: 1997: 37; 2008: 166–71; 2009: 196–9
Briatore, Flavio 42, 104, 160, 173–4, 177–9, 181, 212, 215
Bridgestone tyres 131–2, 134–5, 136–7, 142
British Touring Car Championship 164
broadcasting. see also pit reporting: commentating 44–7, 163–4; live 9, 93–101; post-race media briefings 289; presenting 129–31, 137–8, 141; team orders 84–91, 233–4
Brooks, Lizzie 269, 274
Brown, Zak 223
Brundle, Martin 35, 36, 43, 184; memorable lines 163–4; and Murray Walker 71; and TK 49–50; working style 48, 49–51
Bull, Clive 23–4
Bush, Rupert 31–2
Button, Jenson 185, 191, 192, 193; Brazilian GP 2009: 194–200; joins McLaren 200

Capital Radio 26–9
Carlton Television 32, 34
cars 7–8; aerodynamics 186, 189–90; double diffusers 190–4; explained using cheese 266–7; ground-effect rules 266, 310, 327; McLaren/Ferrari FIA investigations 158–61; safety cars 99, 174, 294, 295, 298, 300–1, 305; safety features 99–100
Chadwick, Jamie 269–70, 274
Channel 4: 24
Chinese GP: trackside fires 8
Chrysalis Sport 31–3
Clear, Jock 62, 191
commentary boxes 45, 48–9
commentating 44–5. see also Walker, Murray; cue cards 46–7; memorable lines 163–4; and talkback 45–6
Cook, Peter 24
cost caps on teams 118
Coughlan, Mike 157–9
Coulthard, David 231, 272
COVID-19: 274–82
Croft, David 163
cue cards 46–7

Daimler team 222
Davidson, Anthony 5–6
Dennis, Ron 14, 58, 62, 71, 189, 212, 214; Alonso/Hamilton rivalry 149–50, 153–4; and Max Mosley 161; McLaren/Ferrari 'spy-gate' controversy 159

di Montezemolo, Luca 203, 240–1, 242
Domenicali, Stefano 15, 158, 240
double diffusers 190–4
Drive to Survive (Netflix) 93–4, 122, 211, 215, 283–6
drivers: characteristics of success 310–13; female 269–74; pressures on 330–1
Duncanson, Neil 33, 81–2

Ecclestone, Bernie 165, 184, 214, 216

F1: cost caps 118; media liaison 12–13; nature of 1–3, 323–4, 330–1; setting up races 123–5; support races 271, 273; travel & accommodation 117–23
F1 Academy 273–4
Ferrari, Enzo 56
Ferrari team. *see also* Bridgestone tyres: 2008 season 165–7, 168–72; cars 165–6, 240; controversial Monaco GP 103–4, 107–11; management style 247–8; popularity with world champions 239–40; and Schumacher 82, 83–4; Spanish GP 1997: 60–1; 'spy-gate' scandal 157–61; team orders controversy 84–91
FIA: Abu Dhabi 2021 investigation 304–5; aerodynamic testing restrictions 186, 189–90; double diffuser appeal 193–4; George Russell investigation 263–4; licensing of circuits 135;

McLaren/Ferrari 'spy-gate' investigations 158–61; Renault 'crash-gate' investigation 176–9; safety car ruling 99, 174, 294, 295, 298, 300–1, 305; single tyre set rule 131–2, 134–5
Fontana, Norberto 61
FOTA (Formula One Teams Association) 188–9
Frentzen, Heinz-Harald 57

Gemini FM 25–6
German GP, 2008: 174–5
Ghosn, Carlos 172, 181
Glock, Timo 164, 168, 169, 170
Goodman, Louise 35, 47, 136
GP2 broadcasting 144–6
Grand Prixs. *see also* individual names: best to visit 126–7; garages 124–5; hospitality buildings 125; mechanics 125; practice sessions 7–10, 48, 49; qualifying 10, 133, 152; setting up 123–5; travel tips 127–8
Grosjean, Romain 100–1, 176, 284
ground-effect rules 266, 310, 327

Haas team: and COVID-19: 275, 277; and Guenther Steiner 284–5; and Mick Schumacher 209–10, 294; Stuart Morrison 260–1
Häkkinen, Mika 22, 61, 62–3, 76
Halo device 99–100
Hamilton, Lewis 16, 235, 304, 308; 2008 season 168–72; 2016 world

championship 314–15; Abu Dhabi
GP 2021: 292–6, 298–9, 304,
308; and Alonso 148–50, 152–7,
159; Australian GP 2020: 276–7;
characteristics of success 310–13;
competitive nature 311; and
Ferrari 239, 310; GP2 career
144–6; and McLaren 146–7; and
Mercedes 220–1; Mission 44
charity 313; and social justice
313–14; and Verstappen 291–308
Haug, Norbert 58
Hedges, David 27
helicopters, medical 9–10
Herbert, Johnny 22–3
Hill, Damon 29–30, 154; and
Jacques Villeneuve 56, 57; and
Schumacher 55, 56
Honda team: double diffusers
189–91; and McLaren 188–9,
244–5; and Mercedes 187; and
Murray Walker 76–7; and Ross
Brawn 15, 185–9
Horner, Christian 215–17; Abu
Dhabi GP 2021: 296–7, 306–7
Hughes, Howard 28–9
Humphrey, Jake 130–1, 184
Hunt, James 41

IndyCar 56, 99, 271
Irvine, Eddie 60
ITV 31–7, 43, 183–4; advert breaks
35–6; *Development Corner*
265–6; *Ted's Notebook* 255,
256–67; *Testing Notebook* 265–6
Jaguar team 216–17, 321

Japanese GP 8; 1996: 40–1; 2011:
235
Jardine, Tony 35, 66, 75
Jordan, Eddie 213–15
Jordan team 213–15
journalism. *see* pit reporting

Kehm, Sabine 105, 108, 209
Kendrick, Jonathan 269, 274
Kovalainen, Heikki 165
Kravitz, Ted 43; Abu Dhabi GP
2021: 300, 304; applies for pit
reporter role 79–81; and the BBC
193, 255–6; and Capital Radio
26–9; and cheese 266–7;
childhood 19–21; and Chrysalis
Sport 31–3, 36–8, 43–4; and
COVID-19: 282; crowd
interactions 167–8; and Dave
Ryan 150–2; *Development
Corner* 265–6; early passion for
F1: 21–3; and Gemini FM 25–6;
and ITV 31–3, 36–8, 43–4,
265–6; *Learn with Lewis* 267;
From the Pit Lane 256;
professional name 25–6; and
Schumacher 105–6; spotter for
ITV 44–50; and student radio
24–5; tape logging for ITV 31–3,
36–8, 43–4; *Ted's Lockdown
Notebook* 283; *Ted's Notebook*
255, 256–67; *Testing Notebook*
265–6; and *Through the Night*
radio programme 23–4
Kubica, Robert 227–8, 229;
accident injuries 249–50;

and BMW Sauber 248; and
Ferrari 248–9; return to F1:
250–1

Latifi, Nicholas 294
Lauda, Niki 16, 221, 222, 319–21
Lazenby, Simon 129–30
LBC *Through the Night* 23–4
Lewis, Dave 31–2
Liberty Media 273
Liuzzi, Tonio 216
live broadcasting 9, 93–101; and
accidents 96–8, 100–1; weather
disrupting 9, 94–6
Lowe, Paddy 221–2

MACh1 32
Malaysian GP 2013: 232–5
Mansell, Nigel 21, 31, 164
Marchionne, Sergio 242–4
Marko, Helmut 215–16, 229
Masi, Michael: Abu Dhabi GP 2021:
292–308; and Jonathan
Wheatley 290–1, 296–7; 'let them
race' philosophy 292; as race
director 289–90; rule breaking in
Abu Dhabi 298–300
Massa, Felipe 89–90, 111; 2008
season 166–7, 168–71, 172, 180–1;
accident in 2009: 203
Mateschitz, Dietrich 216
Mattiacci, Marco 240–2
McKenzie, Lee 184–5, 243, 262, 272
McLaren team 14, 58–9, 223; 2008
season 165–7, 168–72; Alonso/
Hamilton rivalry 147–50, 152–7,

159; cars 165; and COVID-19:
275, 277–8; and Honda 188–9,
244–5; race fixing accusations
61–2; and Renault 245; 'spy-gate'
scandal 157–61
Meadows, Ron 290, 318
Mercedes team 58; Abu Dhabi GP
2021: 300–3, 304–5; buys out
Brawn GP 200, 206; buys out
Honda 187; constructor's
championships 309–10, 317–18;
and hybrid V6 regulations 235;
Spanish GP 1997: 58–60; Spanish
GP 2016: 16; and Toto Wolff
221–2
Miami GP 2022: 253–5
Michelin tyres 131–5, 138, 142
Minagawa, Masayuki 190
Monaco GP 126; 2006: 103–4,
107–10; 2014: 97; 2024: 11
monoculars 13–14
Morrison, Stuart 260–1
Mosley, Max 159–60, 161, 176, 179

Netflix: *Drive to Survive* 93–4, 122,
211, 215, 283–6
Newey, Adrian 58, 262, 302, 318,
327
Nobels, Aurelia 274
Nolan, John 33, 37
Norris, Lando 164, 262, 330–1
Nürburgring 57–8

Park, Richard 26–7
Pérez, Sergio 293, 327
Piper, Kevin 79–80

Piquet, Nelson Jr 42, 173–6, 178, 181
Piquet, Nelson Sr 21, 113, 176, 178
The Piranha Club 214. *see also*
    team bosses
pit lanes 7, 10, 14–15
pit reporting 6–8, 9, 255, 256–60.
    see also Allen, James; Goodman,
    Louise; live broadcasting;
    advice to TK 80–2; face to face
    communication 12; features
    16–17; and media officers 12–13;
    notebooks 16–17, 259–61;
    post-race interviews 14–16,
    90–1, 97, 105–7; practice sessions
    7–10; qualifying 10; and team
    bosses 14–16
pit stops 10–11, 132, 134, 174
Powell, Alice 272
practice sessions 7–10, 48, 49
presenting 129–31, 137–8, 141
press officers 12–13
Pulling, Abbi 274

qualifying 10, 133–4, 152

race control, broadcasting 289–90,
    307
race directors 289–90, 307
race strategy insights 11
radio journalism 23–5
Räikkönen, Kimi 109, 113–14, 165–6,
    201, 279; world championship
    2007: 161–2
Red Bull 218–19; Abu Dhabi GP
    2021: 301, 302, 303–4, 306–7;
    Jaguar takeover 216–17;
    management style 247; and
    Renault 235, 245; and Vettel
    225–7, 229–36
red flag usage 295, 303
Renault team: 2008 season 172–6;
    fixing Singapore 2008 GP 176–81;
    and McLaren 245; and Red Bull
    235, 245
reporting. *see* pit reporting
Ricciardo, Daniel 235, 263
Rider, Steve 32, 33, 131, 149, 164
Roeske, Britta 264
ROKiT 269–70
Rosberg, Nico 16, 206, 314–16
Rosenthal, Jim 34–5, 36, 131, 136,
    137–8, 141
Russell, George 263–4, 328–30
Ryan, Dave 150–2, 171, 272

safety cars 99, 174, 294, 295, 298,
    300–1, 305
Sainz, Carlos 298, 330
Sauber. *see* BMW Sauber team
Schumacher, Michael 55–6, 76,
    112–16, 131–2; and Barrichello
    82–9; controversial Monaco GP
    103–4, 107–11; and Damon Hill
    55, 56; and Ferrari 82, 83–4;
    GP debut 214; and Jacques
    Villeneuve 55–6, 57–8, 60;
    and motorbikes 202; retirement
    113–16, 143, 201–2; retirement,
    returns from 203–6; retirement,
    second 206–7; skiing accident
    207–8; and TK 105–6; and
    Vettel 227

Schumacher, Mick 207, 208–10, 294
Senna, Ayrton 21, 213
Shovlin, Andrew 191, 195, 198, 318
Silver Arrows 200, 206
Singapore GP, 2008: 42, 173–81
Sky Sports 218, 282; *The Notebook* 256–67; *The Qualifying Notebook* 258–60
Smedley, Rob 89–90
Spanish GP 1997: 58–61
Steiner, Guenther 276, 284–5
Stella, Andrea 223
Stepney, Nigel 157–9
Stoddart, Paul 73–5
support races 271, 273
Symonds, Pat 175–6; double diffuser issue 191, 194; Renault 'crash-gate' investigation 178, 180, 181–2

talkback 45–6
tape-logging 32, 33, 37
Taylor, Simon 35, 36
team bosses: competitive nature 211–12; interviewing 14–16; modern day 223; The Piranha Club 214
team orders: Ferrari controversies 84–91; open broadcast of 233–4
television. *see* broadcasting; individual stations
Todt, Jean 85, 86, 108
Toro Rosso 229–31
travel: F1 teams 117–20; spectators 127–8
tyre compounds 10
tyres. *see also* Bridgestone tyres;

Michelin tyres: changing 10–11; single set rule 131–2, 134–5; US Grand Prix 2005 131–42; wet conditions 94–6

United States GP 2005: 131–42
University Radio Exeter 24–5

Vasseur, Fred 15
Verstappen, Jos 325–6
Verstappen, Max 290–1, 307; Abu Dhabi GP 2021: 292–6, 298–9, 304, 306; career 324–9; and George Russell 328–30; and Hamilton 291–308; and Red Bull 325, 327
Vettel, Sebastian 169, 218–19, 279; Abu Dhabi GP 2010: 232; and Aston Martin 237, 282; and BMW Sauber 227–9; as a campaigner 237–8, 266; childhood 226–7; confidence 228–9, 231; and Ferrari 236–7, 246–7, 282; recklessness 226; and Red Bull 225–7, 229–32, 236; retirement 238; and Schumacher 227; and Toro Rosso 229–31; and Webber 230, 233–5
viewers, demographics of 283–4
Villeneuve, Gilles 56
Villeneuve, Jacques 37; and BMW Sauber 227–8; and Damon Hill 56, 57; and Eddie Irvine 60; and Schumacher 55–6, 57–8, 60; Spanish GP 1997: 60–3
Vowles, James 262–3, 294, 318

W Series 269–70; benefits of 270–3;
      financial difficulties 273
Walker, Murray 21, 24, 32, 35, 36,
      46–7, 323–4; advice to TK 81; cue
      cards 46–7; 'Murray-isms' 68,
      163; popularity of 39–42, 43, 54;
      retirement 69–70; and talkback
      45–6; as a travelling companion
      67; tributes to 71–7; warm-up
      routine 51–2; working style 48–9,
      50–4
weather, disruptive 9, 94–6
Webber, Mark 107, 230, 233–5
Wheatley, Jonathan 307; Abu Dhabi
      GP 2021: 290–1, 296–7, 302,
      306–7; and Masi 290–1, 296–7
White, Marcus 25–6
Whiting, Charlie 135, 174, 176, 191,
      194, 288–90
Whitmarsh, Martin 188–9
Whitworth, John 25
Williams, Claire 269, 274
Williams, Frank 57, 212–31
Williams team 61–2, 213
Wirdheim, Bjorn 216
Wittich, Niels 307
Wolff, Susie 273
Wolff, Toto 16, 218, 220–1, 299, 307,
      309–10
women: barriers to F1 participation
      270–1; F1 Academy 273–4;
      sponsorship 274; supported by F1
      teams 273–4; W Series 269–73
World Motor Sport Council 89–90,
      160, 181
Wurz, Alexander 315–16

Acknowledgements

I never planned on writing a book, but when literary agent Kerr MacRae suggested a 'greatest hits' compilation of F1 stories told from my viewpoint, a few ideas came together. That became more ideas when Cassell's Trevor Davies suggested something that takes the reader behind the scenes of F1. So, thanks to Trevor and Kerr for the idea and the team at Octopus: Mel Four, Erin Brown, Karen Baker, Jennifer Veall, Peter Hunt, Chris Stone and Scarlet Furness for guiding a first timer through the editing process. Hopefully this book has achieved most of its brief and for that, I'm indebted to my media centre, pit lane and paddock colleague Adam Cooper, for keeping the project on the track and out of the gravel trap.

I've been doing my day job full-time since 2002 after Neil Duncanson was brave and/or stupid enough to put an F1 obsessive with a face for radio on-air and into the pit lane. For that, I will be forever grateful. The Chrysalis Sport production team taught me many things – how to make great TV, how to put out a mixing desk that's on fire and how the best feature ideas come from a few Caipirinhas and a two-pence piece. Thanks to Rupert Bush, Dave Lewis, John Nolan, Gerard Lane, Jo Hybert, Malcolm Clinton, Andy Spellman, Andrew De Souza, Tim Breadin, Karen Raphael,

Alan Hurndall, Valerie Garford, Sarah Needham, Anneliese Unitt and to Sally Blower for keeping the show on the road. Our lives were always enriched when we were joined by the Anglia Television power trio of Kevin Piper, Kevin Brown and Mr Steve Aldous. The ITV technical team, the exemplary camera ops Andy Parr, Mat Bryant and Keith Wilson, editors Chris Fells, Mark Jakeman and Dave Boyd Moss, comms Rob Walker, Bill Fievez, Tony Kennedy, Dave Hill and the much-missed Neil Crowland plus the graphics OGs John and Jeremy Tidy.

At ITV Network, thanks to the Controllers of Sport Mark Sharman and the great Brian Barwick for his pep talks. I'm indebted to the magnificent Murray Walker, to James Allen, who set the standard for telling stories from the pit lane, to my first co-reporter Louise Goodman, thanks for teaching me the subtleties of F1, and to the incomparable Steve Rider, 'the silver fox', thank you for your generous advice.

At BBC Sport, I'm indebted to Mark Wilkin for his wise editorship, to Barbara Slater and Ben Gallop at TV Centre and on the ground, to Andrew Benson, Sarah Holt, Sunil Patel, Tim Boyd, Richard Gort, Holly Samos, Jason Swales, Anne Somerset, Kay Satterley and Tom Gent. Thanks to the insanely talented editors Robin Nurse and Chris Denton and I loved working with the team of Lee McKenzie, Jonathan Legard, David Coulthard, Jake Humphrey and the irrepressible, irreplaceable Eddie Jordan. For the last decade and more, thanks to my friends and colleagues on the Sky Sports F1 team brought together by Martin Turner, whose encouragement and creativity always shone through. Huge thanks to Stephen Van Rooyen and Rob Webster for their leadership and impeccable judgement. And to the current Sky team, Billy McGinty, Yath Gangakumaran, Jonathan Licht and the unwavering support of Sky Group's Dana Strong. Thanks to my on-air colleagues, Simon Lazenby, Natalie Pinkham, whose sense of fun and charm has lit up our coverage, and Georgie Ainslie (née Thompson), whose mastery of the presenter's art saw me through my nerves at many a studio show. I began working with David Croft when we were both at the BBC and it's a pleasure to continue to listen to the voice of F1 in my headphones every week. Thanks to commentary colleagues Harry Benjamin, Ben Edwards and Alex Jaques. My two fellow reporters, Rachel Brookes, whose calm journalistic professionalism is complemented perfectly by Craig Slater, a brilliantly unconventional and idiosyncratic performer. I'm always grateful to our driver analysts for indulging my on-air questions. In alphabetical order: Jenson Button, Jamie Chadwick, Karun Chandhok, Anthony Davidson, Paul di Resta, Johnny Herbert, Damon Hill, Danica Patrick, Nico Rosberg, Naomi Schiff and Jacques Villeneuve. Not a racing driver, but I always value the quick intelligence of former engineer Bernie Collins when we're working the grid.

Thanks to our superb production team led by Tommy Herz, Jessica Medland and Jack McShane and to my pit lane technical troops – on cameras: Lee Kukor Morgan, Pete Velluet (a good man to have with you on an escalator), Keiran Startup, John Dalton, Simon Seager and Tom Basciano. On sound Dave Haigh, Tiger Harrigan, Jim Sefton and Carlton Waghorn; in the edit Tim Davis, Nick Cliff, Hugh Lutley and Simon Graham; writing online Sam Johnston, Nigel Chiu and James Galloway; and making the production happen, Laura Budd, Erin Cornwell, Georgia Constantinou, Chrissie Malone, Emma Chapman, Donald Begg, Gordon Roxburgh, Bridget Bremner and Jo Slennett. Thanks also to the directors and gallery team at Sky who cut the pictures, keep us to time, fade up the sound and roll the video.

A special mention for Jennie Gow, my opposite number as BBC Radio's presenter and the only pit lane reporter to regularly join me in the actual pit lane during qualifying and the race. Following a stroke in 2023, Jennie has had to learn to talk and write again. The courage and determination she and her husband Jamie (one of my producers at Sky) have shown have been inspirational.

Which leaves one person who's piloted the F1 TV helicopter for nearly 30 years and that's 'the governor' himself, Martin Brundle. From ITV to the BBC to Sky, it's been an honour to be MB's comm box helper and reporter.

I owe a huge debt of gratitude to every F1 driver, team boss, technical director, engineer or mechanic past and present for answering my questions and to their press officers for their kind and professional collaboration. At Formula 1, thanks to Stefano Domenicali and Liam Parker, while I must reserve a special thanks to Bernie Ecclestone. 'You're not too bad nowadays,' he told me recently, 'since you grew up.'

But my professional life would be pointless without the people I do it for – any viewer, listener or reader, thank you for coming into the pit lane with me every week, and thanks to anyone who asked where I was.

Finally, much love, of course, to my family and friends. Thanks for letting me skip so many Sunday lunches. To my smart, funny, independent, bright, brilliant and beautiful daughters, thank you for not minding too much when I go away. Finally, to my 'executive producer', Kate, without whom this book wouldn't exist. Thanks for putting up with my author's travails and for editing the manuscript so brilliantly. I love what's in your head and cherish the daily joys of your companionship. I only hope my first effort is worthy of a mention on the Book Club Review podcast.

About the Author

Ted Kravitz is an experienced broadcast journalist who has reported on Formula 1 for nearly 30 years. During that time Ted has become something of an institution in the pit lane, working successively for ITV, the BBC, Channel 4 and Sky Sports and reporting for four BAFTA award-winning sports programmes in 2006, 2007, 2008 and 2021.

Ted's insider insights have made his on-the-spot reports essential viewing, while his widely acclaimed *Ted's Notebook* programme has become a fan favourite, explaining complex technical details for seasoned followers and taking new audiences behind the scenes with all the news from the sharp end of Formula 1. When not at the race track, Ted flies light aircraft. This is his first book.

Picture Credits

RAISING READERS
Books Build Bright Futures

Dear Reader,

We'd love your attention for one more page to tell you about the crisis in children's reading, and what we can all do.

Studies have shown that reading for fun is the **single biggest predictor of a child's future life chances** – more than family circumstance, parents' educational background or income. It improves academic results, mental health, wealth, communication skills, ambition and happiness.[1]

The number of children reading for fun is in rapid decline. Young people have a lot of competition for their time. In 2024, 1 in 10 children and young people in the UK aged 5 to 18 did not own a single book at home.[2]

Hachette works extensively with schools, libraries and literacy charities, but here are some ways we can all raise more readers:

- Reading to children for just 10 minutes a day makes a difference
- Don't give up if children aren't regular readers – there will be books for them!
- Visit bookshops and libraries to get recommendations
- Encourage them to listen to audiobooks
- Support school libraries
- Give books as gifts

There's a lot more information about how to encourage children to read on our website: **www.RaisingReaders.co.uk**

Thank you for reading.

hachette
LIVRE

---

[1] OECD, '21st-Century Readers: Developing Literacy Skills in a Digital World', 2021,
https://www.oecd.org/en/publications/21st-century-readers_a83d84cb-en.html

[2] National Literacy Trust, 'Book Ownership in 2024', November 2024,
https://literacytrust.org.uk/research-services/research-reports/book-ownership-in-2024